T0184511

Food Sovereignty and Urban Agriculture

/

This book analyses the interplay of urban agriculture and food sovereignty through the innovative lens of the "critical urban food perspective". It focuses on the mobilisation of urban food producers as a powerful response to highly exclusionary dynamics in the agri-food system including insufficient food access and disastrous land dispossessions.

This volume particularly aims to fill the gap in the current literature by engaging with food sovereignty discourses and movements in urban areas. Related activism of urban food producers in the Global South remains underrepresented in practice and in literature. Therefore, this book engages with the lived realities of an urban agriculture initiative in George, South Africa. Building on theoretical notions of the "right to the city" and "everyday forms of resistance", the book illuminates how deprived food producers expose inequalities and propose alternatives. The findings of in-depth empirical research reveal that dwellers perceive farming as a mean to overcome historical segregation, high food prices, and unhealthy nutrition. Hence, they breathe life into food sovereignty in practice and suggest further alliances beyond the city.

The book will be of interest to scholars and students of alternative food politics, agrarian transformation, and food movements as well as rural–urban intersections.

Anne Siebert is a postdoctoral researcher and lecturer at the Institute of Development Research and Development Policy (IEE), Ruhr-University Bochum, Germany. She obtained a joint PhD degree in International Development Studies from the IEE and the Institute of Social Studies, Erasmus University Rotterdam, The Netherlands. Her research experience and interest revolve around food politics and social movements, and how these have shaped dominant agri-food systems, governance, rural–urban interlinkages, as well as research methodology.

Critical Food Studies

Series editors: Michael K. Goodman, *University of Reading, UK*, and Colin Sage, *Independent Scholar*

The study of food has seldom been more pressing or prescient. From the intensifying globalisation of food, a world-wide food crisis and the continuing inequalities of its production and consumption, to food's exploding media presence, and its growing re-connections to places and people through 'alternative food movements', this series promotes critical explorations of contemporary food cultures and politics. Building on previous but disparate scholarship, its overall aims are to develop innovative and theoretical lenses and empirical material in order to contribute to – but also begin to more fully delineate – the confines and confluences of an agenda of critical food research and writing.

Of particular concern are original theoretical and empirical treatments of the materialisations of food politics, meanings and representations, the shifting political economies and ecologies of food production and consumption and the growing transgressions between alternative and corporatist food networks.

Food System Transformations
Social Movements, Local Economies, Collaborative Networks
Edited by Cordula Kropp, Irene Antoni-Komar, and Colin Sage

Metaphor, Sustainability, Transformation
Transdisciplinary Perspectives
Edited by Ian Hughes, Edmond Byrne, Gerard Mullally, and Colin Sage

Food and Cooking on Early Television in Europe
Impact on Postwar Foodways
Edited by Ana Tominc

Hunger and Postcolonial Writing
Muzna Rahman

Food Sovereignty and Urban Agriculture
Concepts, Politics, and Practice in South Africa
Anne Siebert

For more information about this series, please visit: www.routledge.com/Critical-Food-Studies/book-series/CFS

Food Sovereignty and Urban Agriculture

Concepts, Politics, and Practice in South Africa

Anne Siebert

Routledge
Taylor & Francis Group

LONDON AND NEW YORK

First published 2023
by Routledge
4 Park Square, Milton Park, Abingdon, Oxon OX14 4RN

and by Routledge
605 Third Avenue, New York, NY 10158

Routledge is an imprint of the Taylor & Francis Group, an informa business

British Library Cataloguing-in-Publication Data
A catalogue record for this book is available from the British Library

Library of Congress Cataloging-in-Publication Data
A catalog record has been requested for this book

ISBN: 978-1-032-02269-7 (hbk)
ISBN: 978-1-032-02270-3 (pbk)
ISBN: 978-1-003-18263-4 (ebk)

DOI: 10.4324/9781003182634

Typeset in Bembo
by SPi Technologies India Pvt Ltd (Straive)

To my parents

Contents

Acknowledgements

This book bears the imprint of many people. The rich experience, insights, and support of many individuals and institutions made this study possible. I am extremely grateful to Jun Borras; you introduced me to critical agrarian studies and a vital network of critically engaged scholars. Your enthusiasm and thoughtful inputs contributed richly to my "intellectual-political project". Many thanks to Pierre Thielbörger. You encouraged me to do research on food sovereignty in urban settings. I highly appreciate your support in many essential steps of my project.

Thanks to the European Commission, particularly the Erasmus Mundus Partnerships Programme EUSA_ID, for the funding of an extensive research and teaching period at the Institute for Social Development, University of the Western Cape, South Africa. This grant made the research possible. I owe many thanks to the Research School PLUS of the Ruhr University Bochum, which funded additional conference trips and provided the opportunity to present my findings to international audiences, defend and refine my arguments.

Carrying out fieldwork in South Africa would never have been possible without support from the DST-NRF Centre of Excellence in Food Security at the University of the Western Cape, South Africa. A special thanks to Julian May for facilitating the institutional support. Thanks to Raymond Auerbach of the Nelson Mandela Metropolitan University for introducing me to "Kos en Fynbos" and providing expertise during my first visits. Many thanks to Hannah Posern who started earlier with her research on Kos en Fynbos and shared key insights with me. I am deeply indebted to the communities in George for their openness and participation which made this research possible and inspired me in many ways. I would also like to thank all the other organisations and government departments which shared important information with me. Without this assistance it would have been difficult to understand the vast urban agriculture and food sovereignty landscape in South Africa.

In Bochum, many people in and around the Institute for Development Policy and Development Research (IEE) have supported me in various ways throughout these years. First and foremost, many thanks to Wilhelm Löwenstein and Gabriele Bäcker; you encouraged me to pursue a PhD at the IEE and the International Institute of Social Studies (ISS). Your continued support has been an important part of my academic formation since my early days at the Ruhr University Bochum. Many thanks to Christina Seeger, Ruth Knoblich, Stefan Buchholz, Steven Engler, and Tobias Thürer. You have been great colleagues; your friendship motivated me and showed me many new opportunities. Thanks to all the old and new faces at IEE who keep the institute going and growing.

At the ISS in The Hague, the "village" community truly inspired my work and made this sometimes arduous journey enjoyable. I am particularly grateful to Adwoa Yeboah Gyapong, Melek Mutioglu Özkesen, and Yukari Sekine for their positive and intellectual spirit. Thanks to all those who contributed to the incredibly serious but also fun working atmosphere in office 3.33: Britta Holzberg, Raffael Beier, and Toktam Ahnaee.

I am deeply grateful to the good old Giessen-crew, you are never forgotten; your knowledge and continued support motivated me in many ways. Thanks especially to Bridgit Fastrich, Carolin Beinroth, Franziska Ollendorf, and Jonas Metzger. Thanks to my friends and critical thinkers in South Africa: Carolin Gomulia, Giulia Coletti, Kerstin Fischer, and Lena Opfermann. Thanks to Alessa Heuser and all those who engage with food sovereignty in Bochum and the Ruhr area.

I would like to thank Jong-Hoo for joining me early in this journey and for keeping me on my toes. Your love helped me to write this dissertation. I am grateful to my family, who supported me in so many ways, from fresh home-made food to joining me in my stays abroad. To my parents, thank you for always believing in me. To Hannes, for discussing the world and its art with me. To my grandparents and grandmother, for being delightful spirits. And to all the other indispensable ones in my tiny hometown and beyond its boundaries.

1 Why food sovereignty in the city matters now

Under the banner of food sovereignty diverse, mostly small-scale food providers around the world fight for autonomy, invest in ecological diversity, or gain new agricultural knowledge. Exclusionary conditions of agri-food systems[1] have propelled food sovereignty as an alternative concept, social movement, academic discourse, and tacit practice for social and ecological transformation in diverse parts of the world. This is also evident in South Africa, where almost 67% of the population were living in cities, and about 30% were unemployed, in 2019 (United Nations Population Division 2019b; Statistics South Africa 2020). About 10% of the households were vulnerable to hunger[2] in the same year (Statistics South Africa 2019a). With the ongoing COVID-19 pandemic, numbers are estimated to be much higher especially in informal urban areas. Partly in response to these conditions, food production in cities has become both a vital self-help strategy and a governmental intervention to address far-reaching neoliberal ills over the last decade. Driven by these dynamics, urban food initiatives represent a breeding ground for efforts to achieve food sovereignty, including demands for affordable nutritious food and access to land. The urban agriculture initiative "Kos en Fynbos" (KEF)[3] in George fits into this transition and allows to explore new trajectories of resistance rooted in deprived urban settings. The group's practices of localised food provision, community solidarity, and resistance inspire this book in exploring the connection of urban agriculture and food sovereignty in and beyond South Africa.

Confronting the dominant agri-food system

In the global interplay of recurring food, climate, energy, environmental, and health crises, the issue of food scarcity is regaining traction in broader struggles over resource access and control. These developments are occurring in the context of an increasingly globalised agri-food system, which is characterised by expanding industrialisation of the agrarian sector, commercialisation of food production, distribution, and consumption including trade liberalisation in food. Several scholars agree that powerful corporations have strongly shaped food and agriculture around the world since the 1980s (Friedmann 2016, 673; McMichael 2016). This can be traced back to the structural adjustment programmes of the IMF and the World

DOI: 10.4324/9781003182634-1

Bank in the 1980s (McMichael 2013, 41), which initiated the reduction of protective tariffs. In the 1990s, multilateral trade agreements further fuelled the liberalisation of agricultural trade and investments. These interventions facilitated an increase of high-value export-oriented agriculture in the countries of the Global South (Clapp 2012, 60–63; Hall 2015, 408). In turn, countries in the north have been exporting grains to the developing regions and thus constructing dependency in basic food provision (Bernstein 2016, 632; Clapp 2012, 63–74). The transnational corporations' dominant role in food production, processing, distribution, and retail implies that these stages have become highly industrialised, commercialised, and capitalised (Hall 2015, 408). A drive to commodify food, water, seeds, and land is another defining feature of the current system (Siebert 2014, 15–17; Thielbörger 2014, 145–151). Concentrated landownership, restricted food choices, rising food prices, limited market access for smallholders,[4] exploitation of resources, and a need to introduce food aid have been some of the central negative consequences (Hall 2015, 406–407; McMichael 2016, 656). Many African countries have already been deeply integrated into the globalised marketplace. This relates to what Moyo, Jha, and Yeros describe as the "new scramble for Africa"; they particularly point to "the integration of Africa's agriculture and natural resources into world markets, leading to the intensification of labour exploitation and depletion of natural resources" (2019, 3–4). Consequently, insufficient access to labour markets and fights over land, water, food, and other resources challenge adequate livelihoods continent-wide.

Overall, marginalised population groups face harsh conditions, from soaring food prices to land dispossessions. In this context of global capitalism, food plays a central role in both the disintegration and the formation of neglected groups. Recently, Sage, Kropp, and Antoni-Kumar argued that a so-called growing second generation of food initiatives is underway fighting against corporate control around the globe beyond their local communities (2020, 2). Hence, using food as an analytical lens helps to observe dynamics of power relations, injustices, lived experiences, and popular resistance in diverse places and across different scales.

In fact, alternatives to the dominant agri-food system have gained momentum: striking examples include the ever-expanding food sovereignty discourse and food sovereignty movements. The global spread of the food sovereignty discourse can be observed in widening transnational agrarian movements, nascent incorporation into national and international politics, and increasing support by scholar-activists. To date, the construction of food sovereignty discourses has been largely rural and peasant-based; at the same time, urban settings reveal a plethora of concrete food sovereignty actions. With more than 50% of the world population living in cities, urban agriculture has become critically important in local food and income provision (United Nations Population Division 2019a). Activists and scholars therefore call for the domain of food sovereignty to be expanded to include urban food production and people (e.g. Figueroa 2015; McMichael 2016). There is a growing awareness of the proliferation of (un)organised urban agriculture initiatives, yet many of these initiatives are not linked to food sovereignty movements and discourses.

Urban agriculture, exclusion, and calls for food sovereignty

The existing urban agriculture landscape globally is clearly eclectic and emerges from diverse motivations (Eizenberg 2019; Tornaghi 2017). Urban agriculture typically covers an array of food production and consumption processes, comprising, for instance, small-scale urban food production, community-assisted agriculture, school gardens, backyard gardening, farmers' markets, and food collectives. Citizens with highly diverse backgrounds are producing food, ranging from small-scale, peasant-like farming to commercial rooftop farms and affluent people's front door gardens. Many countries have seen a significant increase in community action networks that are organising around food production, distribution, and use; many of these are an indispensable response to food insecurity, malnutrition,[5] and poverty (Andrée et al. 2020; Goodman et al. 2012). Particularly COVID-19 is a reminder of the importance of localised food provision.

The case of South Africa makes the role of urban food production particularly visible in the context of insecure livelihoods. Diverse dynamics of semi-proletarianisation have been observed, which implies that people have no other choice than straddling labour and land-based livelihoods (Jacobs 2018). In the contemporary period, Zhan and Scully emphasise that "the role of land in providing a source of livelihood security has become increasingly important as wage work in the city has been rendered more insecure under neoliberal rules" (2018, 1020). Thus, many South Africans are confronted with tremendous pressures on social reproduction (Cousins et al. 2018, 1063), ranging from inadequate health-care to food insecurity. Laslett and Brenner define social reproduction as "activities and attitudes, behaviours and emotions directly involved in the maintenance of life on a daily basis" (1989, 382). These considerations build the link to the corporate agri-food system, which restricts people in food consumption and production choices (see Chapter 2). Both are increasingly determined by the market. Globally, people struggle in diverse ways to access healthy food or well-located arable land in cities (e.g. Corcoran, Kettle, and O'Callaghan 2017; Gillespie 2016). To over-come these challenges, many seek to intensify labour in various (often informal) occupations and rely on cheap food, self-provision, or community networks (cf. Moyo 2007, 84; Shivji 2017, 10–11). Hence, increasing capitalist control in essential fields of social reproduction forces people to seek alternatives (Tornaghi 2017, 791).

Food cultivation in cities is clearly of highly contested and fragmented nature, which fuels the underestimation of urban food providers' political agency. Hence, it is essential to investigate the background of urban food movements and to deal explicitly with producers' roots, activism, and political engagement (Borras, Franco, and Suárez 2015; Eizenberg 2019). Urban agriculture could be an important source of political momentum in food sovereignty advocacy. It is therefore key to explore the implications and possible meanings of food sovereignty particularly in urban settings.

Food sovereignty was brought to the public debate at the World Food Summit in 1996 by the international peasant movement La Vía Campesina.[6] It concerns "the right of peoples to healthy and culturally appropriate food produced through sustainable methods and their right to define their own food and agriculture systems" (Nyéléni 2007, 1). Since the 1990s, the overarching and increasingly powerful notion of food sovereignty as being both a globalising vision and a political project for a just and democratic agri-food system has spread to various places and settings (Alonso-Fradejas et al. 2015, 433). In general, food sovereignty comprises alternatives to industrial food production and practices fostering ecological and social sustainability (cf. Robbins 2015). Calls for agrarian reform and democratic participation are at the heart of the movement. Evidently, not every kind of production in the city might fit under the banner of food sovereignty. Hence, small-scale, community and bottom-up initiated agriculture and related political values are the focus of this work. Food sovereignty certainly presents a fertile ground for many who are left behind in today's agri-food system – not only peasants in the Global South – to articulate their needs and claim their rights (Akram-Lodhi 2015, 567). In revaluing food provisioning and addressing social inequalities, food sovereignty inspires many urban initiatives. The motivations of urban food producers might build on concerns related to the concentration of corporate power, access to healthy and affordable foods, and the dismantling of racialised food systems (Edelman et al. 2014, 919).

Today, food sovereignty research investigates different layers: the global level (global governance), national legislation and policies, and grassroots and civil society initiatives. On the local level, food sovereignty might be integrated into diverse approaches and practices, creating solidarity economies, smallholder-consumer networks, ecologically sustainable agriculture, and free seed and data banks. According to the six principles of food sovereignty defined in the Declaration of Nyéléni, it focuses on food for people, values food providers, localises food systems, puts control locally, builds knowledge and skills, and works with nature (Nyéléni 2007). In this context, food sovereignty advances two fronts in parallel. First, critical voices, mainly from civil society, expose exclusionary dynamics in the dominant food system. Second, food sovereignty implies creating and proposing viable alternatives to the dominant agri-food system. Various efforts towards food sovereignty are in place and many actors contribute without even being aware of doing so, which can be characterised as the challenge of "putting food sovereignty into practice" and theoretically conceptualising it (Patel 2010, 189; Schiavoni 2017, 2). So far, large transnational agrarian movements including La Vía Campesina have critically shaped the idea of food sovereignty. However, due to their rapid expansion, these movements tend to struggle in reflecting the interests of their members, causing efforts grounded in local realities, such as the everyday struggle of food producers, to be underrepresented (Borras, Edelman, and Kay 2008, 186; Li 2015, 206). However, urban activism might also gain inspiration from larger and international agrarian movements and could benefit from being their ally (Borras 2016, 22; Borras, Franco, and Suárez 2015, 602).

Until recently, scholars' and activists' food sovereignty discourse mainly centred on alternative food production in the rural Global South and often shape a rather rural paradigm (Alonso-Fradejas et al. 2015, 436; Bowness and Wittman 2020). However, urban food initiatives and engaged academics brought the food sovereignty concept and practices to cities in the Global North (e.g. Desmarais, Clays, and Trauger 2017; Figueroa 2015; Roman-Alcalá 2018). Besides food sovereignty's role in practice, the lack of explicit food sovereignty language was highlighted in some cases (Clendenning, Dressler, and Richards 2016, 165; Figueroa

2015, 510). Simultaneously, the rather progressive concept of food justice is often used by urban food movements (Holt-Giménez and Shattuck 2011). It is mainly concerned with structural discrimination and pushes for community empowerment and local consumption. Clendenning, Dressler, and Richards highlight that the focus often remains on rather short-term interventions in the domestic context (2016, 175). Food sovereignty basically comprises the notion of food justice, and it can be argued that both are increasingly interconnected. The food sovereignty discourse goes much further and engages with issues of meaningful citizenship, democracy, and interaction with the state. These aspects and food sovereignty's inherent structural critique as well as its increasing consideration in the Global South are ideally suited to inspire and guide this writing.

This book intends to contribute to rather incipient research on food sovereignty in cities in the Global South. A contribution by García-Sempere et al. (2018) is a case in point, carving out 30 food sovereignty indicators inspired by the work of urban food producers in Mexico. Despite its important methodological contribution, however, the paper by García-Sempere et al. abstains from dealing explicitly with producers' roots, activism, and political engagement. Given the highly fragmented nature of urban agriculture, efforts towards food sovereignty in urban food production might be unexpected and be considered as "quiet" (Visser et al. 2015). In line with these considerations, Borras and others emphasise the importance of investigating the background of urban food movements and their political agenda (Borras 2016, 20; Borras, Franco, and Suárez 2015, 602).

From a scientific and theoretical stance, the diverse practices and contributions to food sovereignty make it challenging for scholars to neatly define it (Schiavoni 2017, 2; Patel 2010, 189). Food sovereignty in discourse and practice has been framed by the six pillars defined in the Declaration of Nyéléni in 2007. However, it is important to keep in mind that food sovereignty is a process and a concept in evolution. For example, Patel maintains that "the struggle for food sovereignty in South Africa, faced as it is with the rebranded but barely reconstructed historical conditions of apartheid,[7] needs to look different than the food fight in France. The Californian Peoples' Grocery in Oakland is unlikely to look the same in another US city" (2005, 82). Consequently, Martiniello urges that food sovereignty needs to be understood: "from the everyday peasant practices which embody an incredibly valuable, historically constructed and territorially grounded knowledge. […] Such insights need to be put into dialogue with global formulations elaborated by transnational agrarian movements" (2015, 521). This book is following this call to focus on local, everyday practices particularly in urban food production, which is still relatively invisible on the ever-expanding food sovereignty radar. Inspired by this perceived silence around food sovereignty attempts in urban and peri-urban settings including the political dimension of urban agriculture, the book introduces a new theoretical lens, namely the *critical urban food perspective*.

While La Vía Campesina literally envisages a "peasant way" and has strong rural roots, food sovereignty's urban front needs to be further explored. In critical agrarian studies, there is a key assumption from Marxist orthodox tradition that peasants are destined to disappear (e.g. Hobsbawm 1994). It might be argued that an "urban peasantry" – which is persisting or is being revived under today's neoliberal developments – can be seen as evidence of the adaptability and resilience of peasants and working people, partly defying this prediction. It is often assumed that "urban farmers have agricultural skills and sensibilities brought from the rural area" (Battersby and Marshak 2013, 454). Hovorka calls it a "sociocultural identity tied to agrarian traditions" (2008, 95). However, as exemplified in a case from South Africa, it is not always the first generation of urban–rural migrants that is active in urban farming (Metzger, Ollendorf, and Siebert 2015). Trying first to make a living as wage-labourers or through other income activities, urban agriculture only becomes an option if other livelihood strategies fail or are restricted. Therefore, so-called agroecological wisdom is prone to get lost in the process of proletarianisation and with the division of labour (Schneider and McMichael 2010, 479–480). This knowledge could potentially be re-assessed in the city.

Van der Ploeg ponders that "peasant-like ways of farming often exist as practices without theoretical representation. Hence, they cannot be properly understood, which normally fuels the conclusion that they do not exist or that they are, at best, some irrelevant anomaly" (2008, 16). When looking at the identities of those growing food and raising livestock in cities, many are involved in diverse and changing livelihood strategies. Different practices of urban food producers, ranging from land occupations to food sharing initiatives, demonstrate that the social base of the food sovereignty movement and practice expands beyond the activism of a revitalised peasantry or the deprived; urban forces can be found in educated middle classes as well as in the deprived working class. Certainly, these groups differ in their socio-economic background; they might contribute to the diverse claims for food sovereignty in the city. In sum, those initiatives have the potential to re-establish local food systems and demand a change in government policies. These responses "from below" have dimensions that cut across class and social identities (e.g. gender, generation, race, and religion) (Borras et al. 2018, 1229). This shows that there is a need for flexible and integrative definitions to define potential food sovereignty actors. Thus, fluid categories such as Jacob's formulation of "urban proletariat with peasant characteristics" (2018) or Shivji's "working people" (2017) are helpful and will be elucidated further in this book.

The aim is to explore how such groups of urban food producers become active in exposing and fighting inequality. Understanding their lived realities, mobilisation, and politicisation helps to shed light on the multidimensional impacts of the dominant agri-food system on the ground. The book zooms in on the case of the urban agriculture initiative KEF which is rooted in the deprived working class in George, a secondary city in the Western Cape, South Africa. While urban agriculture initiatives have the potential to intervene in the unequal agri-food landscape, this raises the question of the meaning of food sovereignty as a mobilising frame and uniting element for alliances. This book asks the central questions of *why and how urban agriculture initiatives emerge, and how do they engage with food sovereignty in the contemporary context of highly exclusionary dynamics in cities and the agri-food system.* It engages with two main research gaps. First, urban food producers' responses and resistances in the Global South remain underestimated politically and underrepresented academically. Second, the relative absence of research on food sovereignty both in discourse and in practice in urban settings in the Global South calls for further investigation.

This book is strongly inspired by the KEF initiative in George. Particularly, the rural–urban setting with diverse small-scale farming activities at the fringes of the city makes the initiative unique. George is a so-called commercial hub in the area and, like other urban areas in the region, has been attracting many migrants from rural areas, particularly the Eastern Cape province and neighbouring countries. However, unskilled labour and low educational attainment are challenges for the full integration in the labour market. In addition, the legacy of apartheid, particularly in terms of spatial segregation and socio-economic exclusion, can still be felt like in many urban areas in the country. In response to these wider inequalities, the group was formed in 2012 and attempts to make healthy nutrition widely available. According to a member, it comprises about 700 deprived backyard gardeners and farmers engaged in small-scale, ecological food production mainly in backyard gardens, in some cases on fallow farmland and institutional land (i.e. school, hospitals, and crèches). It encourages small-scale, ecological food production. On the whole, the social basis of the group consists of small cultivators and workers including many retrenched and un(der)employed dwellers. A large number of members indicated that food cultivation was essential to meet their dietary needs. These realities at the margins show many similarities to other (peri-)urban areas in the country. The focus of the book is on the coloured[8] communities of Blanco and Pacaltsdorp as well as the black community in the largest township[9] of Thembalethu, which are all located at the fringes of the city. George's roots in farming are still reflected in its contemporary spatial composition. Outside the city centre, the municipality comprises small rural agricultural settlements. With the continued growth of the city, rather rural settlements like Thembalethu, which has its roots in the apartheid era, and previously independent farming villages, for instance Pacaltsdorp and Blanco with large swathes of open fields surrounding them, became part of George (Siebert and May 2016). KEF members produce food for self-consumption, while only some sell their produce informally. Besides everyday practices of food production, the movement organises workshops and regular meetings to exchange knowledge, experiences, and the literal fruits of their labours. In sum, members create solidarity networks and strengthen self-reliance. As part of the daily farming practices, the movement demands access to land and healthier food which has caused tensions with the municipality.

It is argued that the initiative engages in change from below by advocating for and contributing to healthier nutrition, improved land access, and heightened community participation. These intentions and interventions are considered as resistance and show parallels to the food sovereignty concept. It was vital that the members and supporters of KEF were keen to share their first-hand experiences and knowledge. Furthermore, their work occurred at an interesting intersection of cooperation and confrontation with several other actors, for instance, the municipality. The unique characteristics of the case study, which combine rural agrarian and urban features, contributed to the development the innovative analytical framework, the critical urban food perspective. This is sketched out and developed further in the following.

Analytical framework: critical urban food perspective

Deprived food producers' harsh realities at the fringes of ever-growing cities and their resistance against diverse exclusions inspire the innovative combination of different theoretical components in a novel critical urban food perspective. The very characteristics of KEF unite features of the city, such as reliance on the food retail sector and urban employment, and features of the countryside including food production as a self-help strategy. In many regions of the world, (rural) agrarian life and livelihoods have been adapted in the context of social change and economic restructuring in close relation to the dominant agri-food system, which causes wider dynamics of rural–urban migration. At the same time, contemporary cities lack benign living conditions for everyone. Insufficient access to shelter, food, and the labour market are just some of the challenges. In this context, some dwellers turn (back) to farming for self-provision and to make a living. Simultaneously, these realities have the potential to spur political activism, silent and out-spoken resistance which build the linkage to food sovereignty movements and actions. Research in such contexts certainly requires critical engagement with the overall conditions and impacts of exclusive agri-food systems and corporate-neoliberal agendas. These conditions and specifically the context of South Africa are discussed in more detail in the following chapter.

While the outlined dynamics are becoming more common globally, the investigation of neglected urban farmers' experiences and struggles call for a hybrid concept that deals with both the urban and rural-agrarian side of the phenomena and therefore ties together related concepts and literatures. So far, many existing theoretical approaches rather tend to engage with the "rural" and the "urban" in isolated boxes. Hence, the critical urban food perspective combines concepts which are rooted in "critical agrarian studies" (CAS)[10] and "critical urban theory" (CUT).

It is apparent that contemporary questions of land and livelihood go beyond an artificial rural–urban divide. Hart and Sitas are concerned about a tendency to research "elements in isolation: 'the land question', the 'labour question', or the 'question of livelihoods' (usually meaning non formal-employment)" (2004, 32; see also Du Toit 2018, 1098; Edelman and Wolford 2017, 960). It is in this context that scholars call for research across disciplines and emphasise urban–rural linkages (Borras et al. 2018; Harvey and Wachsmuth 2012). Bernstein even criticises the food sovereignty discourse for not exploring the interface between the urban and rural contexts (2014, 1052). This book sets out to further advance these points in theory and practice. For this purpose, the proposed critical urban food perspective is to be understood as a reflexive conceptual exploration, a lens through which to investigate specific lived realities while considering dynamically changing contextual factors.

Social and economic exclusion have become close companions in growing cities globally (Brenner, Marcuse, and Mayer 2012). Proponents of CUT argue that the capitalist state and the capitalist class determine urban realties; their actions are firmly related to and propelled by the proliferation of free market forces and a capitalist world economy (Brenner 2019; Harvey 2018; Purcell 2002). This can be observed in increasing real estate investments, restricted land access, a decrease in affordable housing, long working days with insufficient wages, and privatisation of social services and infrastructure. Consequently, choices of ordinary citizens are narrowing, and many are pushed further to the margins (Purcell 2002, 99). The work of critical urban theorists sets out to comprehensively understand how exclusionary dynamics play out in everyday life in contemporary cities. While Marx emphasises that "political economy does not deal with [the worker] in his free time, as a human being" (1963, 76), Lefebvre,[11] one of the pioneers of CUT, intended to fill this gap and engaged with the sphere of social reproduction. It is one of his key assumptions that "workers do not merely have a life in the workplace, they have a social life, family life, political life; they have experiences outside the domain of labour" (Lefebvre 1988, 78). Thus, everyday life goes beyond meeting biophysical needs required to provide adequate labour (Lefebvre 1971, 21). In *Critique of Everyday Life*, Lefebvre shows how the logic of capitalism, described as a "brutally objective power", determines everyday life (1991 [1947], 166). Hence, capitalism implies abstraction of everyday life and its increasing control over social reproduction. Lefebvre's work on the "right to the city"[12] can be understood as a response to this kind of subordination, and several scholars have contributed their contemporary interpretations (e.g. David Harvey, Neil Brenner, Peter Marcuse, Marc Purcell). The right to the city is grounded in everyday life and therefore in struggles of marginalised inhabitants for more democratic urban spaces; it helps to illuminate ways in which urban dwellers expose forms of power, exclusion, and inequality, propose alternatives, and politicise their work (Brenner 2019, 391–392; Marcuse 2009, 194–195).

Beyond Lefebvre's pioneering work, many urban movements have used the right to the city to frame their work, for instance to demand appropriate housing. Recently, it has become an important tool to analyse urban food producers' demands, struggles, and management of space in the so-called food disabling city (Tornaghi 2017). Related current research illuminates how food producers appropriate space according to their needs and wishes away from the hegemony of neoliberal urban agendas and how they subvert the capitalistic logic in everyday social practices (Apostolopoulou and Kotsila 2021; Haderer 2017; Purcell and Tyman 2019). In this regard, this book is inspired by Efrat Eizenberg, Marc Purcell, Laura Shillington, and Chiara Tornaghi. These works facilitate a fresh understanding of the restrictions in social reproduction and alternatives created by people. The right to the city might play out as a demand for food access, which is more than simple affordability. It could be a demand beyond the commodity of food. Food and food production can be considered a means to reclaim space or even seen as a weapon to fight marginalisation in everyday life and thus to claim the right to the city (Granzow and Shields 2019; Shillington 2013; Tornaghi 2014). In fact, the everyday includes peoples' food relations and therefore can serve as a potential source for contestation of larger inequalities (Figueroa 2015, 500).

In South Africa, critical agrarian scholars partly relate the outlined exclusionary developments to a missing linear trajectory from farm to factory, implying that Marx's assumption of full proletarianisation of rural producers is far from reality (see Chapter 1; Du Toit and Neves 2014; Zhan and Scully 2018, 1018). This plays out in growing cites which are characterised by increasing poverty, unemployment, and reliance on social grants (cf. Cousins et al. 2018, 1068–1069). One of the consequences is an ever-expanding fragmented class of labour and working poor (ibid., Du Toit 2018, 1089). Scholars of CAS engage in diverse ways with the rapid transition from rural to urban which has shaped a number of multidimensional crises, for example in terms of food and livelihood insecurity (cf. Edelman and Wolford 2017). Hence, food sovereignty emerged as an important response to inequalities enshrined in the dominant agri-food system. Keeping up with these developments, CAS engage with food sovereignty movements and food sovereignty discourses (e.g. McMichael 2014; Schiavoni 2017; Shattuck, Schiavoni, and VanGelder 2015).

CUT and CAS share a broad Marxist inheritance and often employ a constructive triad of analysis, critique, and search for alternatives. More specifically, both these schools of critical thought are concerned with changing living conditions under capitalism, particularly its consequences for those at the margins, diverse forms of resistance, and the creation of alternatives. By bringing CAS into conversation with CUT, the goal of the critical urban food perspective is to examine how social realities, social reproduction, and space are shaped by capitalism and (re-)constructed by people who maintain relationships to the "soil" in the city. This perspective uses elements of CUT and CAS in the overall frame of the mentioned constructive triad and the right to the city. While Lefebvre's work is one of the inspirations of this book, it can only be the employment and combination of different concepts in the outlined critical perspective which guides research and analysis in the rhythm of contemporary developments and emergent practices. Therefore, the main concepts guiding this work are everyday forms of (peasant) resistance, the notions of the working people and semi-proletarianisation, as well as the production of space. These are key to understand the wider political economy of the prevailing food and agriculture landscape and are introduced further in the following sections and chapters.

While the right to the city departs from exploring everyday life under capitalism in contemporary cities, its initial focus is on exposing inequalities and exclusionary dynamics like lacking access to resources and opportunities and related resistance, transformative demands, and alternatives initiated by the excluded themselves (Madden and Marcuse 2016, 122–124). In this context, it needs to be clarified further who demands rights to the city. Marcuse elucidates: "the demand comes from those directly in want, directly oppressed, those for whom their most immediate needs are not fulfilled: the homeless, the hungry [...] those whose income is below subsistence, those excluded from the benefits of urban life" (2012, 31). Lefebvre locates the revolutionary forces broadly in the working class (1971, 70). But he is also aware that people are often reduced to the working class; following this lead, Purcell and Tyman demonstrate in a case of urban food producers that awareness and narratives reach far beyond their class identity; they point to their interests "as farmers, as community members, and as political activists" and "as communities of colour" (2019, 60–61). From the perspective of agrarian studies, Moyo and Yeros highlight that urban and rural sides "share the same social basis in the semi-proletarianized peasantry, landless proletarians and urban unemployed" (2005, 9), which are part of the periphery of the system. Borras et al.

argue that contemporary capitalism has "radically decreased the level of autonomy and ability of both rural and urban working people to construct or defend their livelihoods" (2018, 1229). Particularly Shivji's concept of the working people provides further guidance to frame these individuals. Departing broadly from loosely connected terms like peasants, rural poor, working people, and labouring people, Shivji unites essential characteristics in the concept of the working people: "the materiality which underlies producers–peasants and pastoralists, proletarians and semi-proletarians, street hawkers [...] – in virtually all sectors is the minimizing of their necessary consumption and maximizing of their labour" (2017, 11). While he abstains from labelling the working people as a "class" in itself or for itself, he is convinced that "it is a 'class' against capital and has great potential in political discourse and mobilization" (ibid. 12). Using these related contemporary thoughts of CUT and CAS, the critical urban food perspective is centred upon the so-called deprived, the working people.

It is in this context that both schools of thought, CAS and CUT, offer related definitions of resistance and political action and highlight the revolutionary potential inherent in everyday life. Gardiner, for instance, points to "envisaging and enacting in different ways of living" which are not determined by capital (2004, 244). On the one hand, it is key to focus on routines, ordinary elements, particularly their repetition, as well as quotidian, mundane existence (Gardiner 2004, 229). On the other hand, the concept of everyday life goes beyond the trivial and embraces the extraordinary, irrational, uncertain, and creative features, which might be connected to the exploration of possibilities and the creation of spaces of experimental utopia as well as negotiated and transitional spaces (Granzow and Jones 2020; Elden 2004, 113). These two opposites often share common elements. The notion of the ordinary, but especially the extraordinary, which contains the signs of non-conformity in everyday life, offers a breeding ground for change and resistance. The related exploration of everyday life sheds light on the realities at the margins. For instance, the analysis of food sources in marginalised communities reveals that many groceries are bought in informal markets or obtained through informal neighbourhood networks.

In a broad way, Bayat frames resistance "from 'dangerous classes' to 'quiet rebels'", which points to overt and covert as well as individual and collective struggles in everyday life (2000, 533). Following this lead, the two rough categories of covert and overt resistance can be used to provide important impulses for analysing the case study. Both have been considered in urban contexts and benefit further from rich conceptual messages and experiences in critical agrarian studies. As highlighted earlier, these resisters are amongst the immediately deprived, the subaltern, the marginalised workers and peasants as well as relatively powerless groups (O'Brien 1996; Roy 2011; Scott 1985, 29). First, covert resistance, what Bayat calls "quiet encroachment of the ordinary", refers to stealthy practices, for instance, to get food, informal jobs, or business opportunities in cities (Bayat 2000, 536; Roy 2011, 228). When exploring resistance in real life it is essential to keep the related consciousness of social action in mind; this is the main difference between resistance and accidental actions or ordinary behaviours (Bayat 2000, 544–545). Bayat and Biekart note that unorganised, individualised attempts have the potential to reap changes on individual level and more broadly (2009, 825). Some refer to this as the Scott type of resistance of the mundane kind (cf. Kerkvliet 2009; van der Ploeg 2008). In this respect, critical urban theory (Bayat 2000, 541; Samara, He, and Chen 2013, 289) was inspired by critical agrarian studies, specifically Scott's study of everyday forms of (peasant) resistance (1976, 1985). Second, resistance in everyday life may also comprise rather overt acts. Here, O'Brien's notion of "rightful resistance" is considered as particularly helpful. In this way, diverse nuances of protests building from already existing rights and values, and how those are used to demand change, can be observed. According to O'Brien, it implies "use of laws, policies and other officially promoted values to defy 'disloyal' political and economic elites; it is a kind of partially sanctioned resistance that uses influential advocates and recognized principles to apply pressure on those in power […] who have not implemented some beneficial measure" (1996, 33). The main differences to notions of hidden resistance are the attention of elites as well as the noisy, open, and public character of overt forms (O'Brien 2013, 1051). Risk of confrontation is reduced by asserting and sticking to core values as well as pressing for accountability (O'Brien and Li 2006, 54 ff.).

Under the banner of the right to the city, resistance and demands for change from below are strongly intertwined with the production of space. Space is relational; it is a social construct and produced through everyday life, through social circumstances, interests and needs; which implies that space is also an instrument of control and power (Eizenberg 2012; Lefebvre 1991 [1974], 26; Purcell and Tyman 2019). In practice, this points for example to the need to recover urban space from forces of economic and political marginalisation (Ghose and Pettygrove 2014, 1108). While these claims often concern land access, Allen and Frediani argue that "urban agriculture might constitute a practice through which marginalised groups might actively claim spaces of daily sociability and political articulation" (2013, 365). For instance, the way in which food is purchased, produced, and consumed is closely connected with the social and natural production of urban space (Shillington 2013, 104).

Lefebvre's conceptualisation of space consists of three interlinked moments: perceived, conceived, and lived space (the so-called spatial triad) (1991 [1974], 38–39). These three moments of space are considered helpful to explore the transformative notions of agri-food movements (cf. Benton-Hite et al. 2017; Eizenberg 2012; Lapina 2017). "Perceived space" (so-called spatial practice) is the actual material space a person encounters; it is concrete, material, physical. It is basically real space – space that is generated and used. It comprises a person's actual experiences of space in everyday life and mundane practices. "Conceived space" points to the representation of space, ideas, and mental construction of space. It is related to abstract conception. This kind of space refers to idealism, measurement, and vision. It is the kind of space produced by planners, architects, or simply by explorers. Bringing in O'Brien's rightful resistance, conceived space also refers to official values, policies, and rights – not all of which have been put into practice. Awareness that those exist could propel citizens' demands to work towards their consideration and implementation. Hence, it is envisioned but not yet lived. Perceived and conceived space – the real and imagined spaces – are blended together in "lived space" and also play out in everyday life (Purcell 2003, 577). Therefore, the different concepts of resistance introduced earlier are used to observe how people change and adapt their lived space according to their needs and wishes. For instance, exclusionary dynamics force citizens to create alternatives, e.g. for income generation and food provision which may be considered as efforts towards food sovereignty.

Many scholars highlight that food sovereignty efforts are grounded in everyday life and in the lived experience of the excluded (Figueroa 2015; Ngcoya and Kumarakulasingam 2017; Shattuck, Schiavoni, and VanGelder 2015). Similar to food sovereignty, the right to the city refers to processes of bottom-up initiated social change, in which marginalised people strive for autogestion (i.e. self-management) including greater direct democracy (Elden 2004, 226; Purcell and Tyman 2019, 60).

With these considerations as point of departure, this book uses two analytical blocks mainly in Chapters 3 and 4. First, building from dynamics of de-agrarianisation and semi-proletarianisation (see Chapter 2), Chapter 3 engages with the notion of everyday life and related exclusionary dynamics. In this regard, the concept of the working people is ideally suited to frame deprived conditions in social reproduction. Second, Chapter 4 zooms in on resistance. In this context, the concepts of everyday forms of (peasant) resistance and the production of space further enrich the analysis. The combination of these concepts in the critical urban food perspective is considered as an open-ended dialectic which is essential to explore people's efforts towards food sovereignty in urban settings. The observation of food sovereignty is further guided by the six food sovereignty pillars defined in the Declaration of Nyéléni.[13]

Food sovereignty gained popularity as a fluid concept, which is adapted in different settings and shaped by different actors seeking an alternative, people-centred agri-food system (Schiavoni 2017; Wittman, Desmarais, and Wiebe 2010). This serves as a reminder for a process-oriented investigation, which focuses on the roots and transformative notions developed by the people. The well-known definition of food sovereignty introduced in the Declaration of Nyéléni provides a point of departure for food sovereignty research. Specifically, the Declaration identifies six pillars[14] which indicate that food sovereignty:

1. focuses on food for people,
2. values food providers,
3. localises food systems,
4. puts control locally,
5. builds knowledge and skills,
6. and works with nature
                    (Nyéléni 2007, 1).

Alimi et al. consider the pillars as "sites and frameworks of interchanges, communication, bargaining, and negotiation" (2015, 56). This book uses them as guideposts, which implies openness to the everyday realities of the people and the way these realities are shaping efforts towards food sovereignty. In the following, a brief overview including first information on the South African agri-food system is provided.

Pillar 1 *Food for the people* refers to the "right to sufficient, healthy and culturally appropriate food for all individuals, peoples and communities [...] and rejects the proposition that food is just another commodity" (Nyéléni 2007, 1). In South Africa, powerful food industries are strongly shaping the food consumer environment by offering processed food, making healthy food unaffordable to the poor, and therefore contributing to mounting health concerns (Faber and Drimie 2016).

Pillar 2 *Values food providers* implies respecting the rights of all people, "who cultivate, grow, harvest and process food; and rejects those policies, actions and programmes that undervalue them, threaten their livelihoods and eliminate them" (Nyéléni 2007, 1). In South Africa, historical tensions, food supply chain formalisation, and increasing commercialisation of agriculture push small food providers further to the margins including those in urban areas (Greenberg 2017).

Pillar 3 *Localises food systems* emphasises the importance of putting "providers and consumers at the centre of decision-making on food issues, [...] protects consumers from poor quality and unhealthy food [...] and resists governance structures, agreements and practices that depend on [...] unaccountable corporations" (Nyéléni 2007, 1). The localisation component is evident in several pillars. In this context, Robbins indicates that localisation can "be viewed in opposition to the wider industrial agriculture model" (2015, 452) – which implies agroecological, peasant, and small-scale agriculture. South Africa's corporate agri-food system is barely in favour of these farming practices.

Pillar 4 *Puts control locally* refers to "control over territory, land, grazing, water, seeds, livestock and fish populations on local food providers and respects their rights" (Nyéléni 2007, 1). Control over the means of food production is the primary focus here. In South Africa, the question of land access is equally unsettled in rural and urban settings (Hendricks, Ntsebeza, and Helliker 2013), while access to water for food cultivation presents another challenge.

Pillar 5 *Builds knowledge and skills* puts emphasis on "food providers and their local organisations that conserve, develop and manage localised food production and harvesting systems, developing appropriate research systems to support this" (Nyéléni 2007, 1). Agricultural knowledge might range from recognising good soil conditions to diverse seed saving practices. South Africa's agricultural policies are in favour of mono-cropping and GMO seeds and thus contribute to the loss of traditional knowledge, seeds, and indigenous vegetables.

Pillar 6 *Works with nature* points to "the contributions of nature in diverse, low external input agroecological production and harvesting methods that maximise the contribution of ecosystems and improve resilience" (Nyéléni 2007, 1). This implies the rejection of harmful agricultural practices. Although small-scale farming following the principles of agroecology does exist in South Africa, this is a niche in an agri-food system characterised by commercialisation.

Marcuse's work on CUT and the right to the city offers an ideal frame to structure the outlined concepts and guideposts within the critical urban food perspective. Such a framing helps to analyse the work of deprived food producers related initiatives and to explore in which ways they challenge the dominant food system and capitalist city. He outlines a process with three intertwined steps:

"expose, propose, and politicise" (Marcuse 2009, 2012). Different forms of resistance and notions of the production of space can be incorporated. Following the considerations above, it can be argued that food sovereignty is operating on two fronts in parallel: critical voices (e.g. urban agriculture initiatives) (1) expose and dismantle exclusionary dynamics in the agri-food system and (2) propose and build viable alternatives. Both fuel the third step, the political dimension. The three steps are framed in the following way:

(1) *Expose* refers to making problems and challenges visible, uncovering their roots (Marcuse 2009, 194); it implies revealing injustice and exclusion in the agri-food system. This comprises the conception and perception of space, and to identifying the best angles of attack and manoeuvring between different tactical stances. In relation to O'Brien's rightful resistance, "exposing" refers to realising which rights, political promises and values are weakly implemented. This implies a desire for change. Thus, what this change looks like and how it can be realised is to be identified. The related analysis of the case study is mainly provided in Chapter 3.

(2) *Propose* refers to adjusting and shaping values, imaginaries, and demands. This requires identifying shared experiences and concerns. The right to the city therefore refers to the creation of a new (future) city, which includes the component of proposing alternatives and acts of resistance (Madden and Marcuse 2016, 125). It could also relate to demanding certain rights. To put it more concretely, proposals, programmes, targets, and strategies are relevant to move into the desired direction (Marcuse 2009, 194). In terms of space, these steps relate to imaginary and visionary spaces (i.e. conceived) as well as lived space. Chapter 4 illuminates the KEF initiative's attempts in this regard.

(3) *Politicise* builds on the exposed and proposed elements. In general, it points to "political awakening" of urban initiatives (Certomà and Tornaghi 2015, 1127). The political is intertwined with everyday life (Haderer 2017, 65). In this way, decisions on political strategising, including tools and tactics, are required, for instance in day-to-day politics, media outlets, or direct interventions like demonstrations. Politicisation might play out in the lived space. Chapter 5 presents the political dimensions of KEF. Beyond a localised perspective, the third part of this chapter applies a wider political perspective by contrasting food sovereignty efforts with Eric Olin Wright's anti-capitalist strategies (2017, 2019). His Marxist sociologist approach allows to bring different critical considerations of this book together and to give nuance to discussed transformative attempts in the context of the broader social system. Simultaneously, and by using the critical urban food perspective, this reflection points out opportunities in how to advance efforts towards food sovereignty.

In sum, the outlined analytical framework is grounded in the quotidian practices of the so-called deprived in the city, the working people, and intends to illuminate how people direct their energy towards transforming the agri-food system. The framework is based on the concepts of resistance in everyday life, the production of space, and food sovereignty. It seeks to illuminate direct impacts of the dominant

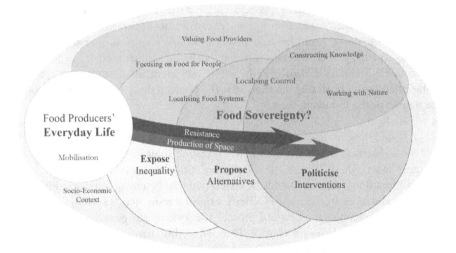

Figure 1.1 Analytical framework: critical urban food perspective. Own compilation.

exclusionary agri-food system. For this purpose, it is uncovered how people expose inequalities. Simultaneously, this book identifies people's proposed alternatives and related politics. This approach is considered as a constructive triad consisting of analysis, critique, and search for alternatives. Certainly, the introduced conceptual perspective offers a toolbox which can be applied elsewhere to understand diverse inequalities in agri-food systems, related forms of resistance and notions of food sovereignty from a viewpoint of those at the margins of society and their allies. An overview of the framework and its different elements is provided in Figure 1.1. The dotted lines framing the different steps and concepts refer to their fluidity.

Research approach

The book is grounded in empirical research. Building from the introduced theoretical considerations, this part introduces the innovative empirical methodology. The research approach is inspired by critical urban theory. In this context, critique is "a means to explore, both in theory and in practice, the possibility of forging alternatives to capitalism" (Brenner 2009, 200). While Brenner highlights the possibility of adjusting theories, he particularly emphasises the need for a flexible and sensitive research methodology. Lefebvre urges researchers to analyse social phenomena in everyday life from two sides, the inductive and the deductive (2002 [1961]). Hence, it is relevant to consider both the wider complex "superstructures" reflected in the existing body of knowledge as well as the emerging chances and events (Lefebvre 1991 [1947], 57). Evidently, research on so-called responses from below requires both deductive (based on theory) and inductive research approaches (based on observations and close engagement with the people on the ground), which allows to develop existing theoretical concepts further. Guided by this advice and by the critical urban food perspective, the attempt was to create powerful empirical accounts of food producers' social realities. This implies uncovering issues and describing behaviours in the context in which they are occurring. These considerations feed into a multifaceted qualitative research approach mainly participatory in nature.

The work of the KEF initiative was the focus of the empirical data collection. This was framed by geographical, historical, and socio-economic information about the area and media coverage of the initiative, for instance, in the local newspaper *George Herald*. The book mainly builds from an intensive fieldwork period in South Africa between January 2016 and March 2017 as well as shorter trips since early 2015. First interactions and exchange with the initiative started in February 2015. For instance, KEF regularly shared insights into their work, including protocols, and reports. Beyond the KEF group, insights from actors shaping the work of the initiative (e.g. local government, non-governmental organisations) and representatives of organisations engaging with food sovereignty on a national level were gathered. The latter was key to frame the overall political landscape in terms of food sovereignty demands in the country. The overall intention was to provide a thick description of the realities in the field; hence, a plethora of field notes were taken, and participant observation was a key means to understand KEF's daily practices and helped to delve deeper into the field. Short informal conversations were integral parts of this method. This included spending time with the food producers in their fields and food gardens to better understand day-to-day procedures. Moreover, KEF gatherings and various stakeholder meetings were attended. Overall, a diverse toolbox of empirical methods was applied, comprising key informant interviews, focus group discussions in combination with neighbourhood walks, and document analysis.

Key informant interviews were mainly conducted with those who were able to and willing to provide insights due to their position and experience. Participants were chosen based on their high level of involvement in the KEF initiative, direct links of their work to KEF, or their organisation's work in the field of food sovereignty. Using purposive snowball sampling, 31 interviews were conducted. To grasp the food sovereignty discourse in South Africa, informal interviews with NGOs which promote food sovereignty on a national level but do not work with KEF (Surplus People Project, Rosa Luxemburg Stiftung, and African Centre for Biodiversity) were conducted. Semi-structured interviews were guided by the overall attempt to understand the interviewees' or organisations' work, motivations, and challenges. In general, the interviews varied in style with some being more formal and others rather being informal conversations. Overall, the names of research participants are kept confidential and in some sections of this book pseudonyms are used.

In addition, focus group discussions combined with neighbourhood walks in the three communities (Blanco, Pacaltsdorp, and Thembalethu) were conducted. These were mainly open gatherings of up to 15 KEF members and were particularly helpful to get familiar with the areas and existing dynamics. The only prerequisite for participation was the willingness to share information openly. The focus was the lived experience regarding food production and their visions. Such rather informal group interviews allow for "joint construction of meaning" (Bryman

2016, 501). These FGDs are part of participatory action research and were inspired by the "Futurehood Methodology" (South African City Futures 2016). Besides understanding the current situation, this approach is thought to encourage people to think about the future, carve out visions, and go beyond the challenges perceived in everyday realities. Part of the approach is a neighbourhood walk in which participants can show places they like and dislike and explain their relation to these places. The intention was to explore places which play an integral role for the participants in daily food consumption and production, e.g. gardens, shops, or markets. In this way, the aim of the discussion and specifically the walk was to motivate participants to share their story.

Document analysis is of complementary nature in this research. Large numbers of documents are referring to the work of KEF and were shared freely by the research participants with the researcher. Particular attention was paid to articles in the *George Herald*, protocols of the initiative's meetings, and governmental reports, mainly of the Local Economic Development Unit, which refer to local urban agriculture related projects. As the *George Herald* has been supporting the movement as a media partner and has regularly published articles about KEF. The movement has been highly appreciative of the *Herald*'s support and regularly contacts the journalists to share stories and photos.

In general, data and findings were mostly combined and triangulated, which helps to represent the reality in an authentic way and, hence, contribute to the validity of the information. In this context, primary material is supplemented by secondary data from a range of documents including media contributions mainly in the local newspaper, governmental reports, and academic writing. In addition, a few sections are supported with publicly available household and consumer data of the government.

One aim of the research was to support the local community to advance their work in developing alternative food environments. The applied research approach is innovative in a sense that it is closely intertwined with the notion of scholar-activism. Being familiar with different initiatives on the ground, "engaged" researchers are interested in "discovering and narrating what these projects are capable of producing" (Certomà and Tornaghi 2015, 1127; Edelman 2009). This involves two issues in this research: accountability and reciprocity. In line with Pulido's thoughts, the attempt was reconcile research interests with that of the community, which might be referred as "bottom-up accountability" (2008, 351; Hall et al. 2017, 4). The intention was to share the initiatives' messages, facilitate interaction between different actors, provide constructive feedback on discussions and everyday interactions with those who participated in this research. A key take-away message from this experience is that academic research has tremendous potential to add value to already ongoing processes of food sovereignty construction, to expand related political space, and to help interlinking different actors advocating for a just agri-food system.

Empirical research poses a number of challenges. The subjectivity of the researcher may limit the way in which the data were gathered and are presented. In general, the strategy a researcher follows is only the manifestation of one particular construction of reality out of multiple possible constructions of social reality from the same data. Using different data sources and combining them (triangulation) is essential in limiting this bias. Furthermore, exchange of thoughts and findings with others familiar with the field helped to ensure that the reality is presented in a meaningful and truthful way. Detailed notetaking and documentation proofed essential to (re)construct the realities authentically at a later stage. Given the language barrier in a few situations, translations from isiXhosa and Afrikaans to English may have interrupted the direct flow of exchange. Furthermore, certain tensions between different actors (KEF and municipality) were observed. The aim in all such cases was to act with sensitivity.

Overview of the book

Departing from the outlined research gaps, the analysis of food sovereignty in this book comprises four chapters. Chapter 2 sheds light on key forces shaping South Africa's commercialised agri-food system to better appreciate the claims for food sovereignty and possible contributions urban agriculture can make. This chapter outlines the nascent food sovereignty discourse at a national level, key proponents, and difficulties in connecting with localised food and agrarian struggles. The final part introduces notions of urban agriculture and its role in exclusionary food landscapes; it particularly zooms-in on the historical and socio-economic conditions in George, Western Cape, which gave rise to the KEF initiative.

The following three chapters – Chapters 3 to 5 – introduce the local perspective in using of the KEF initiative. The analysis mainly applies the concepts of resistance in everyday life, the production of space, and food sovereignty. These specific concepts are illuminated in more detail in the respective chapters. Guided by the book's analytical perspective, the activities of the initiative are clustered in three steps: (1) exposing the inequalities of the prevailing food and agrarian landscape, (2) proposing and creating alternatives, and (3) politicising these interventions. Chapters 3 to 5 are structured along these steps and narrate in detail the lived realities of the movement and sketch out parallels to deprived urban food producers elsewhere. Simultaneously, the relevance and contribution of the proposed critical urban food perspective is emphasised.

Chapter 3 delves into members' socio-economic backgrounds, different types of urban agriculture and the organisational structure. Beyond that, this chapter sheds light on the initial mobilisation to illuminate related inequalities and negative impacts of the commercialised agri-food system on the ground. The realities of the case study mirror diverse conditions of exclusion, e.g. un(der)employment and decreasing availability of healthy food which are evident in growing metropolitan areas in South Africa and beyond. A key message is that agrarian livelihoods are still important despite the overall decline of small-scale farming globally.

Chapter 4 explores food producers' engagement with food sovereignty throughout day-to-day activities and thus the alternatives the group is proposing, for example in terms of informal food access. This chapter is thematically structured by the food sovereignty pillars. Some of the findings exceed the parameters of these pillars, other parts are rather narrow. Consequently, five clusters are framed which provide detailed insights into the lived experience of food production and provision at the fringes of the city, construction of local knowledge, and protection of biodiversity, just to name a few aspects. Chapters 2 and 3 both comprise diverse examples of resistance and production of space.

Chapter 5 engages with the politicisation of the initiative – specific demands and claims – and carves out eight rights to the city which are comprising key elements of food sovereignty. Beyond rather localised efforts towards food sovereignty, this chapter shows parallels to urban food, land, and livelihood struggles elsewhere. Key take-away messages and parallels to community-led initiatives and urban food movement beyond South Africa are highlighted. For this purpose, this chapter emphasises the importance of food sovereignty actions and discourses beyond the rural–urban divide. Moreover, it introduces Eric Olin Wright's anti-capitalist strategies to sketch-out further opportunities and transformative pathways for agri-food movements and their allies.

In Chapter 6, the book concludes that efforts towards food sovereignty in practice and discourse are particularly important in areas in which population groups suffer from diverse exclusions; this may comprise but is not limited to so-called jobless growth, decline of small-scale farming, and broader difficulties in finding benign living conditions in cities and its fringes. The introduced critical urban food perspective is particularly suitable in understanding exclusionary dynamics in everyday life shaped by the dominant agri-food system and related responses and proposed alternatives from below. Indeed, the experience of the urban agriculture initiative in George helps to advance food sovereignty inside and outside of cities and calls for respective social movement alliances. The final chapter of the book summarises key findings, highlights the theoretical contribution, and outlines broader implications for theory and practice.

Notes

1 According to Greenberg, the "agri-food system refers to the full range of procurement, production, distribution, and consumption dynamics within national boundaries, whether formal, regulated and commercial or not" (2017, 957).

2 According to Shisana et al., "Hunger is also referred to as food insecurity, while its absence is considered as evidence of food security" (2013, 145). According to the Food and Agriculture Organization of the United Nations (FAO), "Food security exists when all people, at all times, have physical, social and economic access to sufficient, safe, and nutritious food that meets their dietary needs and food preferences for an active and healthy life. The four pillars are availability, access, utilization, and stability" (2008).

3 "Kos en Fynbos" is Afrikaans, with "Kos" meaning *food* and "en" meaning *and*. Fynbos is a typical heathland vegetation in the Western Cape, which has been adapted to grow in very nutrient-poor and acidic soil. It can easily be incorporated into a garden and is testimony to the little effort needed to start gardening.

4 The term "smallholder" broadly refers to the rural poor and to those farmers who are struggling to make a living. Certainly there are class-based differences: supplementary food producer, allotment holding wage worker, worker peasants, etc. (see further Cousins 2010). Precise class differentiation is not required as part of this study, however, and the term is mostly used to signal the deprived situation of those engaging in farming in the countryside.

5 Malnutrition ranges from extreme hunger to obesity. According to the Western Cape Government Provincial Treasury, "malnutrition (either under- or over nutrition) refers to the condition whereby an individual does not receive adequate amounts or receives excessive amounts of nutrients" (2015, 10). Basically, it can be defined as ill health associated with poor diets.

6 According to Desmarais La Vía Campesina is: "embracing organizations of peasants, small and medium-scale farmers, rural women, farm workers and indigenous agrarian communities in Asia, the Americas, Europe and Africa. It provides opportunities for peasants to articulate a coherent set of demands in the international arena" (2008, 138). In 2016, La Vía Campesina comprised 182 organisations in 81 countries (La Vía Campesina 2018).

7 Apartheid refers to the era of white supremacy in the period 1948 to 1991. South Africans were officially classified according to skin colour, history and language.

8 In South Africa, it is usually distinguished between four different population groups: black (African) (80.7%), coloured (8.8%), Indian/Asian (2.6%), and white (7.9% of the total population) (Statistics South Africa 2019b).

9 People classified as black and coloured were forced to live in townships outside the economic city centres during apartheid. Since the overthrow of the apartheid regime in 1991, the government has been building new residential areas, which have been merging with the original township. These areas have grown massively over recent decades and often incorporate informal settlements.

10 Edelman and Wolford provide a comprehensive definition of CAS including its origins, diverse streams, and contemporary notions: "The roots of today's Critical Agrarian Studies lie in the peasant studies of the 1960s and 1970s, a period when scholars and policymakers [...] viewed the peasantries of Asia, Africa and Latin America as important historical agents" (2017, 962).

11 Lefebvre attempted to fill the gap left by Marx and Engels regarding the realities and problems occurring in cities under capitalism. In this regard, he rejects some of their ideas, and develops others further.

12 The right to the city is not a juridical right (Mayer 2012, 71). Lefebvre applies the term "right" for strategic reasons and in a political sense. It is rather a demand for several new rights, which are carved out through problematising and politicising societal circumstances (Haderer 2017, 64; Marcuse 2009, 192–193).

13 According to the Declaration of Nyéléni: "[Food sovereignty] puts the aspirations and needs of those who produce, distribute and consume food at the heart of food systems and policies rather than the demands of markets and corporations" (2007).

14 These pillars were developed at the Forum for Food Sovereignty in Mali in 2007 and are part of the Declaration of Nyéléni.

References

Akram-Lodhi, A.H. 2015. Accelerating towards food sovereignty. *Third World Quarterly* 36, No. 3, 563–583. DOI: 10.1080/01436597.2015.1002989.

Alimi, E.; Bosi, L.; Demetriou, C. 2015. *The dynamics of radicalization: A relational and comparative perspective.* Oxford: Oxford University Press.

Allen, A.; Frediani, A.A. 2013. Farmers, not gardeners. The making of environmentally just spaces in Accra. *City* 17, No. 3, 365–381. DOI: 10.1080/13604813.2013.796620.

Alonso-Fradejas, A.; Borras, S.M.; Holmes, T.; Holt-Giménez, E.; Robbins, M.J. 2015. Food sovereignty: Convergence and contradictions, conditions and challenges. *Third World Quarterly* 36, No. 3, 431–448. DOI: 10.1080/01436597.2015.1023567.

Andrée, P.; Clark, J.K.; Levkoe, C.Z.; Lowitt, K. 2020. Introduction – Traversing theory and practice. In: Andrée, P., Clark, J.K., Levkoe, C.Z., Lowitt, K. (Eds.): *Civil Society and Social Movements in Food System Governance.* London: Routledge (Routledge studies in food, society and the environment), 1–18.

Apostolopoulou, E.; Kotsila, P. 2021. *Community gardening in Hellinikon as a resistance struggle against neoliberal urbanism: Spatial autogestion and the right to the city in post-crisis Athens, Greece.* Urban Geography. DOI: 10.1080/02723638.2020.1863621

Battersby, J.; Marshak, M. 2013. Growing communities. Integrating the social and economic benefits of urban agriculture in Cape Town. *Urban Forum* 24, 447–461. DOI: 10.1007/s12132-013-9193-1.

Bayat, A. 2000. From "dangerous classes" to quiet rebels'. Politics of the urban subaltern in the Global South. *International Sociology* 15, No. 3, 533–557. DOI: 10.1177/026858000015003005.

Bayat, A.; Biekart, K. 2009. Cities of extremes. *Development and Change* 40, No. 5, 815–825. DOI: 10.1111/j.1467-7660.2009.01584.x.

Benton-Hite, E. et al. 2017. Intersecting race, space, and place through community gardens. *Annals of Anthropological Practice* 41, No. 2, 55–66. DOI: 10.1111/napa.12113.

Bernstein, H. 2014. Food sovereignty via the "peasant way": A sceptical view. *The Journal of Peasant Studies* 41, No. 6, 1031–1063. DOI: 10.1080/03066150.2013.852082.

Bernstein, H. 2016. Agrarian political economy and modern world capitalism. The contributions of food regime analysis. *The Journal of Peasant Studies* 43, No. 3, 611–647. DOI: 10.1080/03066150.2015.1101456.

Borras, S.M. 2016. *Land politics, agrarian movements and scholar-activism*. The Hague: Inaugural Lecture. International Institute of Social Studies.

Borras, S.M.; Edelman, M.; Kay, C. 2008. Transnational agrarian movements: Origins and politics, Campaigns and impact. *Journal of Agrarian Change* 8, No. 2 and 3, 169–204. DOI: 10.1111/j.1471-0366.2008.00167.x.

Borras, S.M.; Franco, J.C.; Suárez, S.M. 2015. Land and food sovereignty. *Third World Quarterly* 36, No. 3, 600–617. DOI: 10.1080/01436597.2015.1029225.

Borras, S.M.; Moreda, T.; Alonso-Fradejas, A.; Brent, Z.W. 2018. Converging social justice issues and movements. Implications for political actions and research. *Third World Quarterly* 39, No. 7, 1227–1246. DOI: 10.1080/01436597.2018.1491301.

Bowness, E.; Wittman, H. 2020 Bringing the city to the country? Responsibility, privilege and urban agrarianism in Metro Vancouver, *The Journal of Peasant Studies*, DOI: 10.1080/03066150.2020.1803842.

Brenner, N. 2009. What is critical urban theory? *City: Analysis of Urban Trends, Culture, Theory, Policy, Action* 13, 198–207. DOI: 10.1080/13604810902996466.

Brenner, N. 2019. *New urban spaces: Urban theory and the scale question*. New York: Oxford University Press.

Brenner, N.; Marcuse, P.; Mayer, M. 2012. Cities for people, not for profit. An introduction. In Brenner, N., Marcuse, P., Mayer, M. (Eds.): *Cities for people, not for profit. Critical urban theory and the right to the city*. London: Routledge, 1–10.

Bryman, A. 2016. *Social research methods*. Oxford: Oxford University Press.

Campesina La Vía 2018. La Vía Campesina Members. https://viacampesina.org/en/wp-content/uploads/sites/2/2018/03/List-of-members.pdf, Accessed 09/01/21.

Certomà, C.; Tornaghi, C. 2015. Political gardening. Transforming cities and political agency. *Local Environment* 20, No. 10, 1123–1131. DOI: 10.1080/13549839.2015.1053724.

Clapp, J. 2012. *Food*. Cambridge: Polity.

Clendenning, J.; Dressler, W.H.; Richards, C. 2016. Food justice or food sovereignty? Understanding the rise of urban food movements in the USA. *Agric Hum Values* 33, No. 1, 165–177. DOI: 10.1007/s10460-015-9625-8.

Corcoran, M.P.; Kettle, P.C.; O'Callaghan, C. 2017. Green shoots in vacant plots? Urban agriculture and austerity in post-crash Ireland. *ACME: An International Journal for Critical Geographies* 16, No. 2, 305–331.

Cousins, B. 2010. What is a "smallholder"? Class-analytic perspectives on small-scale farming and agrarian reform in South Africa. *PLAAS Working Paper Series*, No. 16. http://repository.uwc.ac.za/xmlui/bitstream/handle/10566/4468/wp_16_what_is_a_smallholder_2009.pdf?sequence=1&isAllowed=y, Accessed on 20/04/16.

Cousins, B.; Dubb, A.; Hornby, D.; Mtero, F. 2018. Social reproduction of "classes of labour" in the rural areas of South Africa. Contradictions and contestations. *The Journal of Peasant Studies* 45, No. 5–6, 1060–1085. DOI: 10.1080/03066150.2018.1482876.

Desmarais, A.A. 2008. The power of peasants. Reflections on the meanings of La Vía Campesina. *Journal of Rural Studies* 24, No. 2, 138–149.

Desmarais, A.A.; Clays, P.; Trauger, A. (Eds.) 2017. Public policies for food sovereignty. London: Routledge.

Du Toit, A. 2018. Without the blanket of the land. Agrarian change and biopolitics in post–Apartheid South Africa. *The Journal of Peasant Studies* 45, No. 5–6, 1086–1107. DOI: 10.1080/03066150.2018.1518320.

Du Toit, A.; Neves, D. 2014. The government of poverty and the arts of survival. Mobile and recombinant strategies at the margins of the South African economy. *The Journal of Peasant Studies* 41, No. 5, 833–853. DOI: 10.1080/03066150.2014.894910.

Edelman, M. 2009. Synergies and tensions between rural social movements and professional researchers. *The Journal of Peasant Studies* 36, No. 1, 245–265. DOI: 10.1080/03066150902820313.

Edelman, M.; Wolford, W. 2017. Introduction: Critical agrarian studies in theory and practice. *Antipode* 49, No. 4, 959–976. DOI: 10.1111/anti.12326.

Edelman, M. et al. 2014. Introduction: Critical perspectives on food sovereignty. *The Journal of Peasant Studies* 41, No. 6, 911–931. DOI: 10.1080/03066150.2014.963568.

Eizenberg, E. 2012. Actually existing commons. Three moments of space of community gardens in New York City. *Antipode* 44, No. 3, 764–782. DOI: 10.1111/j.1467-8330.2011.00892.x.

Eizenberg, E. 2019. The foreseen future of urban gardening. In: Certomà, C., Noori, S., Sondermann, M. (Eds.): *Urban gardening and the struggle for social and spatial justice.* Manchester: Manchester University Press, 172–180.

Elden, S. 2004. *Understanding Henri Lefebvre. Theory and the possible.* London: Continuum.

Faber, M.; Drimie, S. 2016. Rising food prices and household food security. *South African Journal of Clinical Nutrition* 29, No. 2, 53–54. DOI: 10.1080/16070658.2016.1216358.

FAO (2008): *An introduction to the basic concepts of food security. Food security information for action: Practical guide.* www.foodsec.org/docs/concepts_guide.pdf. Accessed on 02/12/2020.

Figueroa, M. 2015. Food Sovereignty in everyday life. Toward a peoplecentered approach to food systems. *Globalizations* 12, No. 4, 498–512. DOI: 10.1080/14747731.2015.1005966.

Friedmann, H. 2016. Commentary. Food regime analysis and agrarian questions: Widening the conversation. *The Journal of Peasant Studies* 43, No. 3, 671–692. DOI: 10.1080/03066150.2016.1146254.

García-Sempere, A.; Hidalgo, M.; Morales, H.; Ferguson, B.G.; Nazar-Beutelspacher, A.; Rosset, P. 2018. Urban transition toward food sovereignty. *Globalizations* 15, No. 3, 390–406. DOI: 10.1080/14747731.2018.1424285.

Gardiner, M. 2004. Everyday utopianism. Lefebvre and his critics. *Cultural Studies* 18, No. 2–3, 228–254. DOI: 10.1080/0950238042000203048.

Ghose, R.; Pettygrove, M. 2014. Urban community gardens as spaces of citizenship. *Antipode* 46, No. 4, 1092–1112. DOI: 10.1111/anti.12077.

Gillespie, T. 2016. Accumulation by urban dispossession. Struggles over urban space in Accra, Ghana. *Trans Inst Br Geogr* 41, No. 1, 66–77. DOI: 10.1111/tran.12105.

Goodman, D.; DuPuis, E.M.; Goodman, M.K. 2012. *Alternative food networks: Knowledge, practice, and politics.* Oxon: Routledge.

Granzow, M.; Jones, K.E. 2020. Urban agriculture in the making or gardening as epistemology, *Local Environment*, DOI: 10.1080/13549839.2020.1753666.

Granzow, M.; Shields, R. 2019. Urban agriculture. Food as production of space. In Leary-Owhin, M.E.; McCarthy, J.P. (Eds): *The Routledge handbook of Henri Lefebvre, the city and urban society.* London: Routledge: 287–297. DOI: 10.4324/9781315266589.

Greenberg, S. 2017. Corporate power in the agro-food system and the consumer food environment in South Africa. *The Journal of Peasant Studies* 44, No. 2, 467–496. DOI: 10.1080/03066150.2016.1259223.

Haderer, M. 2017. Recht auf Stadt! Lefebvre, urbaner Aktivismus und kritische Stadtforschung. Eine Rekonstruktion, Interpretation und Kritik. In Kumnig, S., Rosol, M., Exner, A. (Eds.): *Umkämpftes Grün. Zwischen neoliberaler Stadtentwicklung und Stadtgestaltung von unten.* Bielefeld: Transcript Verlag, 63–80.

Hall, D. 2015. The political ecology of international agri-food systems. In Perreault, T.A., Bridge, G., McCarthy, J. (Eds.): *Routledge handbook of political ecology.* London: Routledge (Routledge international handbooks), 406–417.

Hall, R.; Brent, Z.; Franco, J.; Isaacs, M.; Shegro, T. 2017. A toolkit for participatory action research. https://www.tni.org/files/publication-downloads/a_toolkit_for_participatory_action_research.pdf, Accessed on 20/04/21.

Hart, G.; Sitas, A. 2004. Beyond the urban-rural divide: Linking land, labour, and livelihoods. *Transformation*, No. 56, 31–38.

Harvey, D. 2018. Universal alienation. *Journal for Cultural Research* 22, No. 2, 137–150, DOI: 10.1080/14797585.2018.1461350.

Harvey, D.; Wachsmuth, D. 2012. What is to be done? Who the hell is going to do it? In Brenner, N., Marcuse, P., Mayer, M. (Eds.): *Cities for people, not for profit. Critical urban theory and the right to the city.* London: Routledge, 264–274.

Hendricks, F.; Ntsebeza, L.; Helliker, K. 2013. Land questions in South Africa. In Hendricks, F., Ntsebeza, L., Helliker, K. (Eds.): *The promise of land. Undoing a century of dispossession in South Africa.* Auckland Park: Jacana Media, 1–26.

Hobsbawm, E.J. 1994. *Age of extremes. The short twentieth century, 1914–1991.* London: Abacus.

Holt-Giménez, E.; Shattuck, A. 2011. Food crises, food regimes and food movements: Rumblings of reform or tides of transformation? *The Journal of Peasant Studies* 38, No. 1, 109–144. DOI: 10.1080/03066150.2010.538578.

Hovorka, A. 2008. Transspecies urban theory. Chickens in an African city. *Cultural geographies* 15, No. 1, 95–117. DOI: 10.1177/1474474007085784

Jacobs, R. 2018. An urban proletariat with peasant characteristics: Land occupations and livestock raising in South Africa, *The Journal of Peasant Studies* 45, No. 5–6, 884–903. DOI: 10.1080/03066150.2017.1312354.

Kerkvliet, B.J.T. 2009. Everyday politics in peasant societies (and ours). *The Journal of Peasant Studies* 36, No. 1, 227–243. DOI: 10.1080/03066150902820487.

Lapina, L. 2017. "Cultivating integration"? Migrant space-making in urban gardens. *Journal of Intercultural Studies* 38, No. 6, 621–636, DOI: 10.1080/07256868.2017.1386630.

Laslett, B.; Brenner, J. 1989. Gender and social reproduction: Historical perspectives. *Annual Review of Sociology* 15, No. 1, 381–404. DOI: 10.1146/annurev.so.15.080189.002121.

Lefebvre, H. 1971. *Everyday life in the modern world.* London: Allen Lane Penguin Press.

Lefebvre, H. 1988. Toward a leftist cultural politics: Remarks occasioned by the centenary of Marx's death. In Nelson, C., Grossberg, L. (Eds.): *Marxism and the interpretation of culture.* London: Macmillian, 75–88.

Lefebvre, H. 1991 [1947]. *Critique of everyday life Volume I: Introduction.* London: Verso.

Lefebvre, H. 2002 [1961]. *Critique of everyday life. Foundations of for a sociology of the everyday.* Volume II. London: Verso.

Li, T.M. 2015. Can there be food sovereignty here? *The Journal of Peasant Studies* 42, No. 1, 205–211. DOI: 10.1080/03066150.2014.938058.

Madden, D.; Marcuse, P. (2016): *In defense of housing. The politics of crisis.* London: Verso.

Marcuse, P. 2009. From critical urban theory to the right to the city. *City: Analysis of Urban Trends, Culture, Theory, Policy, Action* 13, No. 2–3, 185–196. DOI: 10.1080/13604810902982177.

Marcuse, P. 2012. Whose right(s) to what city? In Brenner, N., Marcuse, P., Mayer, M. (Eds.): *Cities for people, not for profit. Critical urban theory and the right to the city*. London: Routledge, 24–41.

Martiniello, G. 2015. Food sovereignty as praxis. Rethinking the food question in Uganda. *Third World Quarterly* 36, No. 3, 508–525. DOI: 10.1080/01436597.2015.1029233.

Marx, K. 1963. *Early writings*. New York: McGraw-Hill.

Mayer, M. 2012. The "Right to the City" in urban social movements. In Brenner, N., Marcuse, P., Mayer, M. (Eds.): *Cities for people, not for profit. Critical urban theory and the right to the city*. London: Routledge, 63–85.

McMichael, P. 2013. *Food regimes and agrarian questions*. Halifax: Fernwood Publishing (Agrarian Change & Peasant Studies), 2.

McMichael, P. 2014. Historicizing food sovereignty. *The Journal of Peasant Studies* 41, No. 6, 933–957. DOI: 10.1080/03066150.2013.876999.

McMichael, P. 2016. Commentary. Food regime for thought. *The Journal of Peasant Studies* 43, No. 3, 648–670. DOI: 10.1080/03066150.2016.1143816.

Metzger, J.; Ollendorf, F.; Siebert, A. 2015. Smallholder farming under threat? *Digital Development Debates* 16. http://www.digital-development-debates.org/issue-16-food-farming--tradition--smallholderfarming-under-threat.html, Accessed 01/12/20.

Moyo, S. 2007. The land question in southern Africa: A comparative review. In Ntsebeza, L., Hall, R. (Eds.): *The Land Question in South Africa. The challenge of transformation and redistribution*. Cape Town: HSRC Press, 60–84.

Moyo, S.; Jha, P.; Yeros, P. 2019. The scramble for land and natural resources in Africa. In Moyo, S., Jha, P. K., Yeros, P. (Eds.): *Reclaiming Africa. Scramble and resistance in the 21ˢᵗ century*. Singapore: Springer, 3–30.

Moyo, S.; Yeros, P. 2005. The resurgence of rural movements under neoliberalism. In Moyo, S., Yeros, P. (Eds.): *Reclaiming the land. The Resurgence of rural movements in Africa, Asia and Latin America*. London: Zed Books, 8–64.

Ngcoya, M.; Kumarakulasingam, N. 2017. The lived Experience of food Sovereignty. Gender, indigenous crops and small-scale farming in Mtubatuba, South Africa. *J Agrar Change* 17, No. 3, 80–496. DOI: 10.1111/joac.12170.

Nyéléni 2007. Synthesis Report Nyéléni 2007 Forum for food sovereignty. Forum for Food Sovereignty. nyeleni.org/IMG/pdf/31Mar2007NyeleniSynthesisReport-en.pdf, Accessed 09/01/21.

O'Brien, K.J. 1996. Rightful resistance. *World Pol.* 49, No. 1, 31–55.

O'Brien, K.J. 2013. Rightful resistance revisited. *Journal of Peasant Studies* 40, No. 6, 1051–1062. DOI: 10.1080/03066150.2013.821466.

O'Brien, K.J.; Li, L. 2006. *Rightful resistance in rural China*. Cambridge, New York: Cambridge University Press (Cambridge Studies in Contentious Politics).

Patel, R. 2005. Global fascism revolutionary humanism and the ethics of food sovereignty. *Development* 48, No. 2, 79–83. DOI: 10.1057/palgrave.development.1100148.

Patel, R. 2010. What does food sovereignty look like? In Wittman, H., Desmarais, A.A., Wiebe, N. (Eds.): *Food Sovereignty. Reconnecting food, nature and community*. Oakland, California: Fernwood Publishing, 186–196.

Pulido, L. 2008. FAQs. Frequently (un)asked questions about being a scholar activist. In Hale, C.R. (Ed.): *Engaging contradictions. Theory, politics, and methods of activist scholarship*. Berkeley: University of California Press, 341–365.

Purcell, M. 2002. Excavating Lefebvre: The right to the city and its urban politics of the inhabitant. *GeoJournal* 58, No. 2/3, 99–108. DOI: 10.1023/B:GEJO.0000010829.62237.8f.

Purcell, M. 2003. Citizenship and the right to the global city: Reimagining the capitalist world order. *International Journal of Urban and Regional Research* 27, No. 3, 564–590. DOI: 10.1111/1468-2427.00467.

Purcell, M.; Tyman, S.K. 2019. Cultivating food as a right to the city. In Tornaghi, C., Certomà, C. (Eds.): *Urban Gardening as Politics*. Oxon: Routledge, 46–65.

Robbins, M.J. 2015. Exploring the "localisation" dimension of food sovereignty. *Third World Quarterly* 36, No. 3, 449–468. DOI: 10.1080/01436597.2015.1024966.

Roman-Alcalá, A. 2018. (Relative) autonomism, policy currents and the politics of mobilisation for food sovereignty in the United States. The case of Occupy the Farm. *Local Environment* 23, No. 6, 619–634. DOI: 10.1080/13549839.2018.1456516.

Roy, A. 2011. Slumdog Cities. Rethinking subaltern urbanism. *Int J Urban Regional* 35, No. 2, 223–238. DOI: 10.1111/j.1468-2427.2011.01051.x.

Sage, C., Kropp, C. and Antoni-Komar, I. 2020. Grassroots initiatives in food system transformation: the role of food movements in the second "Great Transformation". In: Kropp, C., Antoni-Komar, I., Sage, C. (Eds.): *Food systems transformations. Social movements, local economies, collaborative networks*. London: Routledge, 1–19.

Samara, T.R.; He, S.; Chen, G. 2013. Locating right to the city in the Global South. Abingdon, Oxon: Routledge.

Schiavoni, C.M. 2017. The contested terrain of food sovereignty construction. Toward a historical, relational and interactive approach. *The Journal of Peasant Studies* 44, No. 1, 1–32. 10.1080/03066150.2016.1234455.

Schneider, M.; McMichael, P. 2010. Deepening, and repairing, the metabolic rift. *The Journal of Peasant Studies* 37, No. 3, 461–484. DOI: 10.1080/03066150.2010.494371.

Scott, J.C. 1976. *The moral economy of the peasant. Rebellion and subsistence in Southeast Asia*. N.J.: Yale University Press.

Scott, J.C. 1985. *Weapons of the weak. Everyday forms of peasant resistance*. Princeton. N.J.: Yale University Press.

Shattuck, A.; Schiavoni, C.M.; VanGelder, Z. 2015. Translating the politics of food sovereignty: Digging into contradictions, uncovering new dimensions. *Globalizations* 12, No. 4, 421–433. DOI: 10.1080/14747731.2015.1041243.

Shillington, L.J. 2013. Right to food, right to the city. Household urban agriculture, and socionatural metabolism in Managua, Nicaragua. *Geography, Planning and Environment* 44, 103–111. DOI: 10.1016/j.geoforum.2012.02.006.

Shisana, O.; Labadarios, D.; Rehle, T.; Simbayi, L.; Zuma, K.; Dhansay, A. et al. 2013. *South African National Health and Nutrition Examination Survey (SANHANES-1)*. Cape Town: HSRC Press.

Shivji, I.G. 2017. The concept of "working people". *Agrarian South: Journal of Political Economy* 6, No. 1, 1–13. DOI: 10.1177/2277976017721318.

Siebert, A. 2014. *Die Global Governance des Wassers. Eine Untersuchung der Wasserpolitik und städtischen Versorgungslage in Uganda*. Duisburg-Essen: Working Papers on Development and Global Governance (5). DOI: 10.13140/RG.2.2.36071.85925.

Siebert, A.; May, J. 2016. Urbane Landwirtschaft und das Recht auf Stadt. Theoretische Reflektion und ein Praxisbeispiel aus George, Südafrika. In Engler, S., Stengel, O., Bommert, W. (Eds.): *Regional, innovativ und gesund. Nachhaltige Ernährung als Teil der Großen Transformation*. Göttingen: Vandenhoeck & Ruprecht, 153–168.

South African City Futures 2016. Futurehood methodology. With assistance of South African Cities Network, African Centre for Cities, CSIR, Architects Collective (South Africa), Johannesburg Development Agency, Mandela Bay Development Agency. http://cityfutures.co.za/methodology, Accessed 08/11/20.

Statistics South Africa 2019a. General household survey 2019. http://www.statssa.gov.za/publications/P0318/P03182019.pdf, Accessed on 18/04/2021.

Statistics South Africa 2019b. Mid-year population estimates 2019. https://www.statssa.gov.za/publications/P0302/P03022019.pdf, Accessed on 18/04/2021.

Statistics South Africa 2020. Quarterly Labour Force Survey. Quarter 4: 2019. Pretoria. http://www.statssa.gov.za/publications/P0211/P02114thQuarter2019.pdf, Accessed on 18/04/2021.

Thielbörger, P. 2014. *The right(s) to water.* Berlin, Heidelberg: Springer Berlin Heidelberg.

Tornaghi, C. 2014. Critical geography of urban agriculture. *Progress in Human Geography* 38, No. 4, 551–567. DOI: 10.1177/0309132513512542.

Tornaghi, C. 2017. Urban agriculture in the food-disabling city: (re)defining urban food justice, reimagining a politics of empowerment. *Antipode* 49, No. 3, 781–801. DOI: 10.1111/anti.12291.

United Nations Population Division 2019a. World Urbanization Prospects: The 2018 Revision (ST/ESA/SER.A/420). New York: United Nations. https://population.un.org/wup/Publications/Files/WUP2018-Report.pdf, Accessed on 18/04/2021.

United Nations Population Division 2019b. World Urbanization Prospects: 2018 Revision. South Africa. https://data.worldbank.org/indicator/SP.URB.TOTL.IN.ZS?end=2019&locations=ZA&start=1960, Accessed on 18/04/2021.

van der Ploeg, J.D. 2008. *The new peasantries. Struggles for autonomy and sustainability in an era of empire and globalization.* London: Earthscan.

Visser, O.; Mamonova, N.; Spoor, M.; Nikulin, A. 2015. "Quiet food sovereignty" as food sovereignty without a movement? Insights from postsocialist Russia. *Globalizations* 12, No. 4, 513–528. DOI: 10.1080/14747731.2015.1005968.

Western Cape Government Provincial Treasury 2015. Socio-economic Profile George Municipality. Working Paper. Cape Town. www.westerncape.gov.za/assets/departments/treasury/Documents/Socio-economic-profiles/2016/municipality/Eden-District/wc044_george_2015_sep-lg_profile.pdf, Accessed on 06/02/21.

Wittman, H.; Desmarais, A.A.; Wiebe, N. 2010. The origins & potential of food sovereignty. In Wittman, H., Desmarais, A.A., Wiebe, N. (Eds.): *Food sovereignty. Reconnecting food, nature and community.* Oakland, California: Fernwood Publishing, 1–32.

Wright, E.O. 2017. How to be an anti-capitalist for the 21st century. *THEOMAI Journal Critical Studies about Society and Development* 35, No. 1, 8–21.

Wright, E.O. 2019. *How to be an anticapitalist in the 21st century.* London: Verso.

Zhan, S.; Scully, B. 2018. From South Africa to China. Land, migrant labor and the semi-proletarian thesis revisited. *The Journal of Peasant Studies* 45, No. 5–6, 1018–1038. DOI: 10.1080/03066150.2018.1474458.

2 Rethinking South Africa's agri-food system

Notions of food sovereignty and urban agriculture

South Africa represents a rapidly changing food environment comprising striking malnutrition rates, as well as powerful corporate interests and neoliberal policies (Faber and Drimie 2016; Greenberg 2017). These conditions fuel claims for food sovereignty. The following sections provide insights on the country's co-called commercialised agri-food system to better understand entry points for sovereignty in discourse and practice. In part, a visible and articulated food sovereignty discourse comprising movements for land reform can be observed; this refers mainly to established campaigns and organisations active across provinces including their specific localised projects, for instance the Surplus People Project and the South African Food Sovereignty Campaign. Efforts towards food sovereignty span across macro-level policy discourses to rather micro-level interventions. However, work of local community initiatives presents rather hidden attempts to create alternative food systems on the ground, for instance in deprived urban areas. In this context, it is key to illuminate an ever-growing and diverse landscape of urban and peri-urban agriculture.

The chapter is divided into three parts. First, the country's agri-food system is sketched out. Second, the related food sovereignty discourse, its roots, and central actors are explored. This is supported by interview material. In the last part, notions of urban agriculture are elucidated including an overview of the "Kos en Fynbos" (KEF) initiative, specifically the socio-economic and historical background of the case study location, the city of George in the Western Cape. Hence, this chapter establishes the empirical context.

South Africa's commercialised agri-food system

The country has been well integrated in the globalised and industrialised agri-food system through the political economy of its food and agricultural landscape. For instance, it has been deeply embedded in global agri-export markets and at the same time has become a net importer of various food and agricultural products.[1] Historical particularities including the apartheid legacy and the strong corporate turn have contributed to a comprehensive commercialisation of all steps along the food value chain and simultaneous de-agrarianisation of the country (Sanders et al. 2014, 21). Twenty-seven years after the first democratic elections following minority rule, the intersection of race and class and their connection to food,

DOI: 10.4324/9781003182634-2

water, and land access are still evident (Du Toit 2018; Shisana et al. 2013). In sum, this has amounted to an increasingly unjust agrarian structure and an extremely exclusionary food system, which frame the point of the departure of this book. The following sections provide a sketch of the related dynamics and processes of agrarian change.

Inequality in South Africa's agri-food system has deep roots in the past. Although intense debate over land ownership dates back to colonial times,[2] it was mainly during the early decades of the twentieth century and the apartheid era that millions were displaced from independent production or from employment on farms (Jara and Hall 2009, 209; Lahiff 2009, 103). The Natives Land Act in 1913 allocated only about 7% of the arable land to the black African population (Lanegran and Lanegran 2001, 673). Mass dispossession and relocation mainly to so-called homelands[3] and to townships were the consequences. The act became integral in pathing the way for the apartheid era. In the following decades, moreover, it was part of the political attempt to restrict these farmers in land and livestock size, market access, and any means required to expand their farming activities and livelihoods. Bundy frames these developments as the "decline of the South African peasantry" (1972, 1979). In contrast, white farmers were not confronted with these constraints and profited from governmental support during apartheid. All this played out in a dual system and later in a more specific "two track policy" maintaining black farmers at subsistence level and promoting the interests of white large-scale farmers (Kirsten, van Zyl, and van Rooyen 1994, 20).

Since the 1980s, South Africa's agricultural policies have followed a neoliberal ideology to satisfy and compete in global markets (Sihlongonyane 2005, 145). Moreover, de-regulation of the agricultural sector contributed to corporate control of the agri-food system (Greenberg 2010, 5). With the restructuring in agriculture applying strict rationalisation, the logic of efficiency, profit orientation, and technological advancement, farmers have continued to lose access to land, demand for agricultural labour has declined, and opportunities for small-scale entrants are very limited (Magdoff, Foster, and Buttel 2000, 79–80; Scully and Britwum 2019, 421). The strong corporate turn is also visible in the country's focus on single crops, the related rapid uptake of genetically modified seeds, corporate concentration in the seed sector, and related seed policies (African Centre for Biosafety 2009).[4] At the same time, the commercialised agricultural sector depends heavily on an increased use of chemical fertilisers and pesticides. Not only are these developments perceived as a threat for biodiversity, but small-scale and subsistence farmers tend to be excluded by these profit-driven interventions (Sanders et al. 2014, 30–31). Progressive governmental attempts to strengthen small-scale farmers have been largely ineffective (Aliber and Hall 2012, 548; Greenberg 2010, 13). Thus, the situation of 167,000 smallholder households stands in harsh contrast to that of approximately 40,000 large-scale commercial producers (i.e. those who farm for a main or extra source of income) (Aliber and Mdoda 2015, 18). As Bernstein argues: "South African agriculture and agricultural policy since 1994 has done little, if anything, to 'transform' the circumstances of the dispossessed – rural and urban classes of labour – whose crises of social reproduction remain grounded in the inheritances of racialized inequality" (2013, 23). Although land reform has been a key element of post-apartheid reconciliation policies, it has only benefited a few and the government has been unable to rectify land inequality in both rural and urban areas (Hornby et al. 2017, 3–4).

Du Toit and Neves refer to South Africa's rural underdevelopment and rapid urbanisation as "jobless de-agrarianisation" (2014, 834 and 848). A linear transition from farm to factory is not evident. Jara and Hall describe the dire consequences: "[T]his rural-to-urban migration locates them in zones of continued economic exclusion, structural unemployment and the apartheid legacy of underdevelopment, thus clothing them in the new mantle of the "urban poor"" (2009, 209). High rates of rural to urban migration moreover cause an increase in unskilled labour which tends to keep wages low. Unemployment hits non-white population groups particularly hard. Many people are engaged in informal labour across multiple income activities, or rely on social grants, which all plays out in complex pressures on the social reproduction of "classes of labour" (Cousins et al. 2018, 1061). While almost 30% were unemployed in 2019, about one-fifth of those who were working were engaged in the informal sector (Statistics South Africa 2020). Referring to these developments, further "sprawling [of] low income townships" has been expected (McDonald 2008, 31). These realities at the margins fuel further challenges such as difficulties in food and land access, leading to partial self-provision, for instance in growing food. This ties in with the outlined theoretical considerations in the introduction of this book. It is this specific context, which calls for further attention in the wide food sovereignty landscape.

The country's corporate turn has complicated the situation for small producers, suppliers, and consumers, especially through food supply chain formalisation and concentration since the early 1990s (Du Toit and Neves 2014, 838; Peyton, Moseley, and Battersby 2015, 38). Large-scale multinational food and beverage companies (also referred to as "Big Food" corporations) have been shaping food prices and have been active in transforming dietary patterns by offering highly processed and convenience foods[5] (Greenberg 2017, 468). Particularly, a small number of agribusinesses shape the market and reap high profits from increasing food prices (Jara and Hall 2009, 212). Consequently, consumers with less financial means are restricted in their food choices as healthy diets including fruits and vegetables are associated with higher costs, whereas so-called cheap and nasty food has become more widely available (Faber and Drimie 2016, 53). While rising food prices have hit the poor hard, the informal sector has played an important role both in income and food provision. On the whole, the country presents a "rapidly changing food environment" comprising a sharp contrast between a high prevalence of obesity and simultaneously high numbers of people suffering from hunger (Claasen et al. 2016, 1).

In addition, diverse situations of crises have an impact on food prices. Particularly in the aftermath of the global food crisis in 2007/2008 and during a severe drought in 2015/2016, South Africa experienced steep food price rises which most affected low-income households (Drysdale, Bob, and Moshabela 2021). Food price inflation poses an extra burden on these households which already spend a large amount of their income on food. Due to market openness and deep integration in global trade, shocks in international food prices are directly mirrored on a domestic level (Termeer et al. 2018, 88). More recently, during the COVID-19 pandemic, periods of lockdown including limited operation of informal food traders confronted many deprived South Africans with higher food prices, restricted food availability, and lower incomes (Wegerif 2020, 798). For poor households, food is often the first and only flexible expenditure category in times of insecure livelihoods. Certainly, social grants and alternative means of food provision, e.g. food parcels and soup kitchens, regained relevance. Moreover, climate change is hitting the country hard (cf. Satgar and Cherry 2020, 325). In addition, poorer neighbourhoods were particularly affected by the impacts of intensive periods of drought and related water restrictions which also included shut-offs in household water supply. For low-income households, it is often more difficult to afford bottled water, and overall food preparation at home becomes challenging. South Africa's recurring energy crisis is also relevant here and might reach new heights in the future given the country's strong reliance on imported oil and coal-fired power. Energy shortages complicate food provision, storage, and preparation.

Beyond specific impacts of climate, environmental or health crisis, the state, in its orientation towards the market, seems unable to protect people both in rural and urban areas from the larger inequalities of the commercialised agri-food system, for instance food insecurity. In this regard, Pritchard et al. describe the double face of the state in South Africa's agri-food system, which they refer to as "stepping back and moving in" (2016). On the one hand, the government sticks to policies promoting liberalised economic growth, stressing property ownership and partially deregulated markets, and creating an enabling environment for corporate activities (Greenberg 2017, 490; Pritchard et al. 2016, 704). On the other hand, the state is constitutionally obliged to realise the right to food. The Constitution of the Republic of South Africa 1996 specifically states in Section 27(1): "Everyone has the right to have access to − (b) sufficient food and water". Sub-section (2) adds "The state must take reasonable legislative and other measures, within its available resources, to achieve the progressive realisation of each of these rights" (ibid.). This requires translation into suitable policies. Hence, the government is trying to mitigate income and food poverty through a broad range of social grants, feeding schemes, and food aid (Greenberg 2010, 1; Pritchard et al. 2016, 705). The response of local authorities and development organisations to food insecurity is often embedded in the overall context of development efforts including the promotion of self-help mechanisms like food gardens (Webb 2011, 195). NGOs and community-based organisations have been playing an important role in food provision, but they are not directly in a position to target the systemic roots of the problem. In general, the state thus far failed to introduce integrated policy strategies targeting food insecurity.[6]

The informal sector[7] is still a significant player both in income and food provision (including food retail, distribution, and preparation) and is often neglected by the government (Greenberg 2017, 480; Peyton, Moseley, and Battersby 2015, 37; Skinner and Haysom 2017, 5). Crush highlights that, "the informal food economy comprises a dense and diverse network of informal markets, suppliers, transporters, mobile traders, hawkers, retailers and street food vendors who sprawl across the landscape, making food more accessible and affordable in low-income areas" (2016, 11). In situations of hunger and financial shortages, people turn to informal markets, and advantages like credits, cheaper prices, and smaller package sizes become valuable. Furthermore, residents with lower incomes often rely on alternative food networks, such as food acquired from neighbours or civil society organisations (Haysom 2016, 7). Research in Cape Town's townships shows that, because of their formalised nature, often distant locations, and lower-quality food, supermarkets only have a limited capacity to alleviate food insecurity compared to informal food networks (Paganini, Lemke, and Raimundo 2018, 406; Peyton, Moseley, and Battersby 2015, 37). During COVID-19, the importance of informal food traders in localised, affordable, and fresh food supply became ever more visible. Governmental restrictions of these small-scale operators contributed to food insecurity during the pandemic (Wegerif 2020).

Given a wide range of socio-economic inequalities visible in the agri-food system, a question arises regarding the role of civil society in counteracting these dynamics. Since the late 1990s, the African National Congress (ANC)-led government has been confronted with recurring civil society-led protests for comprehensive service delivery, higher wages, lower food prices, free education, land access, farm worker rights, or anti-GMO campaigns to name just a few (cf. Hart 2013, 28 ff.; Mitlin 2018, 565 ff.). Hart highlights several key events of public protest between two defining moments in Bredell in 2001 and Marikana in 2012 (2013, 28 ff). The case of Bredell refers to a mass land occupation located between Pretoria and Johannesburg and the brutal eviction by government forces. Bredell exposed the land question and contributed to the formation of the Landless People's Movement (ibid. 30). In the case of striking mine workers fighting for higher wages at Marikana, the state resorted to coercive means resulting in the massacre of 34 mine workers in 2012. Hart frames these mobilisations under the banner of "the dialectic of protest and containment" (2013, 28 ff.), referring to the state's abilities to appease responses from below, circumvent civil society's demands, and thus combine coercion and consent.

More recently South Africa's land debate gained new attention with the ANC being challenged from the left. The Economic Freedom Fighters used the land question as a metaphor to question the ruling party's efforts in the realm of economic freedom and more generally in decolonialisation (Kepe and Hall 2018, 128). Hart frames these mobilisations as "popular antagonism" which amounted to a ferocious debate over expropriation without compensation of white-owned land, ANC's plans to amend the constitution, and in part fragmentation of the ANC (2019, 318–319; Kepe and Hall 2018, 129). In early 2021, President Cyril Ramaphosa is optimistic that parliament will approve an amendment of the constitution which implies that land may be expropriated – mainly from white owners – without compensation. The basic intention remains to overcome racially skewed land ownership rooted in colonial periods. Overall, land reform particularly the aspect of expropriation continues to be a key issue in the country's political debates comprising pending questions regarding specific formulations, implementation, and outcomes. Ultimately, meaningful engagement with affected communities and support in efficient land use management remain of utmost importance in this context. Challenges in land access and reform are further illuminated in several chapters of this book. Satgar and Cherry highlight that currently several responses "from below" embracing the notion of food sovereignty can be observed particularly in response to hunger and climate shocks (2020, 327). This kind of resistance continues to grow; in the wake of the COVID-19 pandemic proponents of food sovereignty and food justice have made crucial contributions to the C19 People's Coalition advocating for alternatives to the country's commercialised agri-food system.[8]

Particularly during the last years, several civil society-led struggles on the ground have shown parallels to food sovereignty. Yet food sovereignty seems to be far from being a well-established and galvanising concept for diverse social movements and grassroots' initiatives in South Africa (Ngcoya and Kumarakulasingam 2017, 480). This perceived silence on food sovereignty on the part of civil society can be partly traced back to the country's history of racial segregation (Seekings and Nattrass 2006, 49). Broader campaigns and NGO-led work on food sovereignty, such as the South African Food Sovereignty Campaign or work of the Rosa Luxemburg Stiftung[9], have only recently begun to unite diverse food sovereignty sympathisers. The following section illuminates these nascent efforts and their potential ties to rather isolated struggles for a people-centred agri-food system.

Food sovereignty in discourse: roots, actors, and challenges

Confronting the country's commercialised agri-food system including its apartheid legacy is certainly very challenging and mirrored in the issues of rather fragmented food sovereignty attempts. This part is based on (grey) literature, interviews, and discussions with representatives of the African Centre for Biodiversity (ACB), the Surplus People Project, and the South African Food Sovereignty Dialogue of the Rosa Luxemburg Stiftung Southern Africa. These organisations actively engage with food sovereignty campaigns and use the food sovereignty concept in their work. While SAFSC and the Co-operative and Policy Alternative Center (COPAC) did not respond to interview requests, rich information on their work is publicly available on their websites and in several academic publications. This section is divided into three parts. The first briefly sketches the key difficulties in establishing a wider and more intensive food sovereignty discourse in the country. The second maps the existing food sovereignty attempts, introducing three key land reform alliances and showing how their work has paved the way for today's food sovereignty organisations and alliances. The third part highlights further endeavours and related urban issues.

Initial challenges

South Africa is one of the countries which enshrined the "right to food" in its constitution (see previous section). Despite this, implementation and interventions are only slowly taking shape, and many households still do not have reliable and sufficient access to food. Moreover, food sovereignty has not been adopted as state policy. In contrast, several governments have either explicitly or implicitly adopted food sovereignty into state policies. These include Argentina, Bolivia, Brazil, Ecuador, El Salvador, Guatemala, Indonesia, Mali, Nepal, Nicaragua, Peru, Senegal, and Venezuela (Schiavoni 2017, 2; Wittman 2015, 180; Wittman, Desmarais, and Wiebe 2010, 8). Most of the country's neoliberal food and agricultural policies do not align with the notion of food sovereignty and thus fail to target systemic inequality enshrined in the agri-food system as outlined earlier. Corporate control of the food value chain, reaching from aggressive promotion of GMO seeds to increasing market power of food retail giants, keeps spaces for alternatives narrow and difficult to establish (Satgar and Cherry 2020, 322). Several struggles on the ground, like the ongoing land debate, farm worker strikes, or consumer protests against escalating food prices, do not directly operate under the banner of food sovereignty.

This relative silence on food sovereignty from civil society can be partly traced back to South Africa's history of racial segregation, ethnic fragmentation, and a related culture of separation and discrimination (Bernstein 2011, 46; Levin and Weiner 1996, 100; Siebert 2020, 402). Its agrarian structure is characterised by a historical concentration of land and consequently a highly fragmented smallholder base (Cousins 2010). According to an expert of ACB, "The majority of small black farmers is very marginal to the system. There is not really widespread consciousness [...] around smallholder farming. What we are missing in South Africa is the middle [...] suddenly you jump to 5,000, 10,000 hectare farms".[10] Smallholders can basically be considered as micro-producers with rather tiny pieces of land including backyard gardening. The harsh contrast to these mostly self-subsistent farmers are large-scale, commercial farms (cf. Cousins 2010). This is related to the systemic underdevelopment of black small-scale farming under apartheid and the continued focus on large-scale commercialised agriculture (Kepe and Hall 2018, 132).

In addition, South African farm workers, many of whom work in vineyards and large-scale white commercial farms, miss a greater degree of recognition and representation. In this regard, an expert of the South African Food Sovereignty Dialogue, suggested that "farm workers do not feel represented by rather urban oriented labour unions", which is complicated by the fact that working conditions on farms are barely regulated.[11] Many of these marginalised people – smallholders and farm workers – never experienced political enfranchisement, which relates to their restricted perceptions "as citizens and bearers of rights" (Du Toit 2018, 1089). These might be some of the main reasons why rural communities have been perceived as "invisible" in political struggles (Ntsebeza 2013, 136). Greenberg summarises in this regard: "In South Africa, that [corporate] encroachment took place decades ago and the challenge for the food sovereignty movement in South Africa is to rebuild some kind of small-scale family farm sector that incorporates a critical analysis of the existing agri-food system" (2013, 18). He basically highlights that in most countries where the concept of food sovereignty is proliferating, both in discourse and practice as a movement, the material base is rooted in reliance on and consciousness of small-scale food producers. However, in South Africa this material base is lacking and "has to be constructed along with the discourse" (ibid.). Throughout the interviews, it became clear that this seems to be a common understanding amongst experts. The following sections engage further with these challenges for the food sovereignty movement, and the weak social and material base in the country.

Incipient attempts towards food sovereignty

One might assume that the South African Landless People's Movement (LPM) with its ties to La Vía Campesina and the Brazilian Landless People's Movement (Movimento dos Trabalhadores Rurais Sem Terra – MST) would act as a food sovereignty lighthouse in the country. But its successes were rather short-lived, mainly occurring between 2001 and 2005 (see Ntsebeza 2013, 138–140). However, related alliances have played a critical role in propelling a people-centred land debate and generating impulses to question the corporate agri-food system as a whole. In this context, NGOs such as the Surplus People Project, the COPAC, and the ACB started to advocate for food sovereignty. This work is reflected in the Agrarian Reform for Food Sovereignty Campaign (FSC) and the SAFSC. Moreover, the Rosa Luxemburg Stiftung Southern Africa initiated the South African Food Sovereignty Dialogue in late 2016 with the aim of strengthening exchange between different food sovereignty proponents. Building on discussions with some of these organisations, the food sovereignty landscape, particularly the alliances for land reform and food sovereignty, are outlined below.

Alliances for land reform

In the aftermath of apartheid, land redistribution and land reform became a central feature in reconciliation efforts. This section focuses on the rise and fall of the National Land Committee (NLC), Landless People's Movement (LPM), and Alliance for Land Reform and Agrarian Movements (ALARM). These organisa-tions have been crucial in representing the "people" in land debates and paving the way for public recognition of related deprivations since the 1980s. Their aims have parallels with the food sovereignty mission. Specifically, the LPM – with its official link to La Vía Campesina – connected South African land struggles to global food sovereignty movements and discourses.

In the 1980s, prior to the founding of the LPM, several NGOs and community-based organisations emerged that focused on forced removals in rural areas (Ntsebeza 2013, 136). Many of these joined the National Land Committee which united them in their fight for justice in land access (Hart 2013, 30). In the 1990s, the NLC cooperated with the ANC to advise on land reform. However, the ANC favoured a market-led land reform approach which implies a strong dependence markets including landowners' decisions on what to sell, for which price and when (Hall and Cliffe 2009, 3; Ntsebeza 2013, 137). The NLC's attempts to persuade the ANC to adopt a more progressive direction failed in 1996/1997. Ntsebeza describes the complicated situation in the following way: "On one hand they [committee affiliates] were drawn into implementing the limited [...] land reform programme. Besides this, NGOs were under pressure from their donors to collaborate with the government. These factors contributed to the weakening of NGOs and [...] to an almost total neglect of rural mobilisation" (2013, 138).

Continued struggles for land access and the limitations of the ANC's land reform programme fuelled the formation of the more radical LPM. Grounded in the NGO-based NLC and launched in 2001, the LPM was perceived as significant in the resurgence of resistance in South Africa's land debate: it challenged the government's lack of acknowledgement and limited reparations of land injustices in urban and rural areas (Ntsebeza 2013, 136–137). The formation of the LPM was moreover inspired by land occupations and related forced removals in Bredell in 2001 (see the first part of this chapter). It was LPM's vision to go further than previous land reform attempts, which were strongly driven by the NLC. Mangaliso Khubeka, one of LPM's founders, described LPM's intention "to be the mouthpiece of the people" (cited in Baletti, Johnson, and Wolford 2008, 302). Spurred on by the land reform events in neighbouring Zimbabwe, the LPM became recognised internationally between 2001 and 2003. In this context, "MST activists from Brazil helped to train, educate and inspire South African activists" (Baletti, Johnson, and Wolford 2008, 290). The size of the movement and its wide support became visible at the World Summit on Sustainable Development in Johannesburg in 2002. LPM joined with 5,000 delegates and marched with about 25,000 other landless people and NGOs against neoliberal policies (ibid. 302). In South Africa, this was the first time since the 1950s that a movement with a strong base amongst the landless themselves had gained momentum.

Strongly inspired by the Brazilian MST, LPM opted several times for confrontation through land occupations. In Gauteng, for instance, LPM supported several informal dwellers who faced the threat of forced removals (Greenberg 2004, 17–18). The NLC observed these direct confrontations with the state rather critically, which caused tensions in their cooperation and within the LPM (Ntsebeza 2013, 139). Despite its growing fragmentation, LPM, together with other land reform organisations, successfully pushed for the national Land Summit in 2005. Initiated by the Department of Land Affairs, the intention of the Summit was to pave the way for fast-track land reform. In the preparatory phase of the event, both NLC and LPM suffered from several internal tensions. This led to the formation of ALARM which hoped to unite organisations and social movements struggling for a people-centred land reform.

More than 20 organisations, including the Young Communist League, Trust for Community Outreach and Education (TCOE) and the LPM, were members of the alliance (Hall and Ntsebeza 2007, 16). Several of these groups and organisations had been aligned with the NLC, and the alliance's mission was in "people-centred rural transformation" comprising land transfer and security for people relying on the land for their livelihoods (Ntsebeza 2013, 141). However, during the summit, disputes between ALARM and LPM over their perceptions of the intended nature of land and agrarian reform, which had not been clearly communicated between these groups, became public and raised doubts about the overall intentions in South Africa (Mngxitama 2006, 40). Sihlongonyane described the fragmentation, issues, and confusion between the different organisations: "Whilst instances of collaboration are apparent, and common objectives, programmes and grievances recognizable, there is no clear common enemy. Whereas some attack government at national level, others attack it locally, others at the provincial level, while others blame privatization/globalization" (2005, 158). After the Land Summit, the future of ALARM was rather uncertain. Ntsebeza concluded that "ALARM, as with the Landless People's Movement, faded away. Most of the remaining organisations in the land sector continued to operate in isolation, although now and again they worked together" (2013, 142). Despite the attention it had attracted, the summit's key outcomes remained limited, with a tendency to position land reform within a market framework (i.e. "willing buyer, willing seller" principle) without outlining coherent alternatives (Hall and Ntsebeza 2007, 17).

Today, the LPM is fragmented and still weakened by the tensions amongst its affiliates in its initial phase. This is exemplified in the experience of a former member and local secretary of the LPM, Lekhetho Mtetwa, who resigned and observed past developments particularly close alliances of members with political parties critically (cf. Zabalaza 2019). His doubts can be traced back to the experience with an LPM member who intended to use the movement when running for a municipal position linked to the Democratic Alliance in 2010. Although this did not work out, a rupture of the movement remained and was further complicated in 2014, when several members joined the party of the Economic Freedom Fighters (ibid.). This reveals current problems of political party capture and corruption in the LPM. Today, only some localised efforts in Gauteng and Limpopo remain. In sum, the diverse attempts of LPM, NLC, and ALARM in the past provide a glimpse into the difficulties of cooperating with different organisations and uniting varying interests. Moreover, their experiences show the challenges in generating and maintaining a movement with a strong groundswell character.

Jacobs argues that "the absence of a nationwide land movement in South Africa should not be interpreted as the absence of land struggles. Instead there have been many localized struggles for land" (2018, 16). He refers for instance to a long-term urban land occupation at the fringes of Cape Town, which was supported by the Agrarian Reform for Food Sovereignty Campaign and the affiliated Surplus People Project through direct actions in 2009. Ntsebeza similarly emphasises that several NGOs such as the Surplus People Project and TCOE continue in their attempts to strengthen the social base for land reform (2013, 143). Their work forms part of growing food sovereignty efforts in the country, which are outlined in the following section.

Alliances for food sovereignty

Food sovereignty efforts are incipient in South Africa. An expert from ACB – a member organisation of the South African Food Sovereignty Campaign (SAFSC) and a core member of the African Food Sovereignty Alliance (AFSA) highlights, "The food sovereignty movement [in South Africa] is still a bit premature [...]. Despite the fact that there are a lot of organisations doing work, [...] we are all very fragmented".[12] This implies that different organisations advocating for food sovereignty are operating in pockets and fail to create stronger ties to the people on the ground. During the discussion, it moreover became clear that a disconnection between those NGOs and intellectual people and grassroots struggles on food including food sovereignty issues is visible. The following sections illuminate these observations and introduce the relevant actors in a chronological order.

The Agrarian Reform for Food Sovereignty Campaign, also simply called Food Sovereignty Campaign (FSC), can be considered as one of the highly visible driving forces of the national food sovereignty discourse. The NGO Surplus People Project (SPP) was founded in the 1980s as a response to land evictions in the Western Cape (see previous section) and was the main initiator of FSC. Founded in 2008, the FSC is a member of La Vía Campesina. SPP still seeks agrarian reforms that benefit people oppressed by apartheid and colonialism (Sihlongonyane 2005, 153). Experts from SPP highlighted in a discussion in 2016 that governmental land redistribution and support to farming excludes many people, and these people are in the focus of their work, for instance, poor people, farm workers, farm dwellers, rural people, especially women. This is the target group of SPP's interventions. SPP supports people in their struggle for food sovereignty mainly in terms of land access in the Northern and Western Cape Provinces. SPP started its interventions in urban areas dealing with housing and farming demands of marginalised dwellers. They were active in fighting evictions and demanding alternative land. In this regard, SPP fought several successful court cases for land access for smallholders in the Western and Northern Cape. One example was a land occupation by farmers at the fringes of Cape Town, the SPP expert remembers "They were occupying state land, provincial government land, that belonged to the Department of Human Settlements, and the department wanted to evict them. So, we took the department to court, and then there was a court order that said you had to provide them with alternative land".[13] SPP and FSC helped the farmers to sort out the issue of alternative land. Despite these kinds of successes, the discussion also reveals disappointment. An expert of SPP and FSC ponders: "[we are] helping people to get access to land, water, extension, fighting for the right kind of policies. We are very cynical also about the policy space. I mean for the last 20 years we can't say that we have achieved a lot".[14] However, despite these difficulties, SPP's experiences and support could also be essential for the neglected farmers at the fringes of Thembalethu in George, who are introduced as part of the case study of this book. Many of these farmers are still hoping to benefit from the government's Proactive Land Acquisition Strategy or at least postponed development plans (see Chapters 3 and 4).

The SPP strongly shapes the interventions of the FSC. The FSC has a similar focus, on "small-scale farmers, farm workers/dwellers, landless people, forestry dwellers, farmers with insecure land tenure, rural women, rural youth and rural dwellers in parts of the Western and Northern Cape Provinces" (Food Sovereignty Campaign 2018). The movement has been engaging in various forms of direct actions, including public protests and land occupations (ibid.; Jacobs 2018, 13). An expert of the Rosa Luxemburg Stiftung Southern Africa highlighted that, as well as supporting smallholder-led interventions, the campaign has also been able to shed light on the "deprived situation of farm workers and supported them".[15] Thus, the campaign has been active in connecting rural and urban struggles involving farm workers and landless, which is also mirrored in its past leadership by an urban worker-farmer, who occupied and produced food on land at the fringes of Cape Town. Jacobs elucidates: "early on it [the FSC] recognized the importance of breaking the artificial divisions between rural and urban, small-scale farmers and farm workers – divisions that some movements in Africa are unable to overcome" (2013, 6). Hence, it is certainly one of the achievements of the FSC to start recognising and supporting urban food sovereignty struggles.

Despite its countrywide recognition, the campaign's activities are very localised in the two provinces, they are barely represented on the internet and thus are difficult for outsiders to trace. When I spoke with experts of SPP, I did not know details about the land struggles at the fringes of George (as part of the case study), but they were interested to get more information and to link up with the farmers, which certainly can be a stepping stone for an alliance.

In part, some of the shortcomings of the FSC have been addressed by the younger SAFSC. Founded in 2015, this publicly highly visible umbrella organisation has argued that food sovereignty is a necessary response to "the corporate-controlled food system" and therefore critically engages with diverse issues ranging from the country's seed and land politics to increasing food prices and declining nutrition (South African Food Sovereignty Campaign 2015b). COPAC, an NGO, initiated the SAFSC and has acted as the secretariat of the campaign since its foundation. Formed in 1999, COPAC envisions a grassroots solidarity economy; in this context, it presses for food sovereignty, worker cooperatives, and socially owned renewable energy to name just a few key working areas.

Over recent years, the campaign has been slowly forming as a national exchange platform with 55 member organisations; according to the campaign's statement, it "emerges out of a need to unite organisations, social movements, small scale farmers, farmworkers and NGOs championing food sovereignty into a national platform in advancing food sovereignty strategically in South Africa" (South African Food Sovereignty Campaign 2015a). One of the members is the earlier mentioned ACB. It started as the African Centre for Biosafety with a strong focus on rejecting genetically modified organisms (GMOs) and later broadened its work. Established in 2003, this research and advocacy organisation became well-known continent-wide for its work on seed sovereignty and agricultural diversity which relate to resilience in the context of food sovereignty. Some of the campaign's members, like ACB, are part of AFSA. AFSA is a Pan African platform comprising NGOs, regional producer organisations, and social movements launched in 2011. Its formation was a direct response to the new Green Revolution in Africa (Holt-Giménez and Shattuck 2011, 134).

The SAFSC gained wider attention through public actions ranging from protests against bread corporations to boycotts of GMO seeds (South African Food Sovereignty Campaign 2015b). Moreover, already in its early phase, SAFSC developed a proposal for a Food Sovereignty Act, "which would aim at creating the conditions for food sovereignty and mandate the state to undertake certain actions to support this" (Drimie and Pereira 2016, 19). It was in 2018 that the SAFSC, in cooperation with the Friedrich Ebert Stiftung, published the Peoples' Food Sovereignty Act. It was developed in a "participatory process" comprising several workshops, dialogues with grassroots organisations, and presentations to government departments and the parliament (South African Food Sovereignty Campaign and Friedrich Ebert Stiftung 2018, 3). The People's Act is intended to serve as a "campaigning tool" and "will be further developed" (ibid., 2). Its chapters cover a wide range of issues from seed to trade policies with the objective "to change the laws governing the food system" (ibid. 4). Endeavours like these exemplify the campaign's desire to influence high-level structures, including government policies (Williams and Satgar 2020, 276). It is assumed that these attempts contributed to new points of view by the South African Human Rights Commission regarding the realisation of the right to food; they recognised food sovereignty in addition to food security as an important approach (Satgar and Cherry 2020). Overall, Satgar and Cherry highlight that the SAFSC is transforming dominant structures through four forms of power: (1) *symbolic power* through actions like seed banking and bee keeping, (2) *direct power* through methods like the bread march and visits to the government to hand over the People's Food Sovereignty Act, (3) *movement power* in uniting different actors, sharing capacities and deepening so-called knowledge commons, and (4) *structural power* through organisations like local markets or schools which promote practices in line with food sovereignty (2020, 330–332).

Still, the SAFSC faces diverse challenges. So far, it has been mainly promoted and advanced by high-profile NGOs and well-established community organisations. Grassroots voices are virtually absent in the SAFSC. According to an expert of ACB, "the important thing about the food sovereignty campaign is to have some base amongst the poor of this country".[16] While the campaign has been striving for inclusion and representation of different actors, also in its national coordinating committee and in meetings, the direct involvement of grassroots participants, especially small farmers, has remained a challenge (Cherry 2016, 120–121). The interview partners, whose organisations collaborate with the campaign, pointed to several weaknesses in the SAFSC. Two major concerns can be summarised as follows:

(1) *Not a naturally grown network and representation of civil society/social movements remains weak:* This refers to the lack of a material and social base amongst the deprived (e.g. smallholders, consumers, farm workers). In this context, an expert from SPP pointed out that "people can speak for themselves", referring to the people on the ground as well as involved organisations.[17] This chimes with the thoughts of the expert from Rosa Luxemburg Stiftung Southern Africa, who moreover considers the foundation of the campaign as "top-down" and "not a network which was initiated in different corners".[18]

(2) *Challenge to connect all actors and involve small organisations which lack capacities:* The interviewees referred for instance to organisational limitations, "the geographical challenge, how to interlink people, [...] how to bring them together physically";[19] "we [ACB] are like 5 people [...] so, we have to limit what we can do".[20] In addition, most of the meetings and events take place in Johannesburg, where COPAC is based, and Skype conversations are not a suitable alternative for several community-based organisations (cf. Cherry 2016, 103). The expert from Rosa Luxemburg Stiftung Southern Africa highlighted, "COPAC holds a strong lead, because no one else can do it".[21]

These mostly organisational aspects are not to be understood as an overall critique. The interview partners also reflected critically on their own work. In sum, it was strongly emphasised that networks like the SAFSC are important, but a base amongst the poor and the marginalised is essential to truly make a difference. Beyond these considerations, it was for example suggested to strengthen farmer-to-farmer and farm-worker networks in terms of knowledge exchange, but also to combine the provision of technical, organisational, and political knowledge and to make it applicable in practice. According to the expert from ACB: "people who got political awareness must transfer it to the farmers and there is a tendency to rely too heavily on rhetoric and not enough on the practicality of the issue. [...] Sharing information and technical support is not good enough".[22]

Beyond these critical aspects highlighted in the interviews, the work of the SAFSC is widely welcomed. For instance, Drimie and Pereira stress that the South African food system is in crisis; they refer to food insecurity, increasing consolidation of the food value chain, and the decline of smallholder farmers to name just a few aspects of that crisis (2016, 2). At the same time, they are in favour of alternative approaches in line with the food sovereignty concept, e.g. promoting nutritious food, eating communally, and reasserting cultural identity linked to food. In this context, they emphasise: "Although still building momentum, the Food Sovereignty Campaign could provide an important impetus for supporting and building such approaches. The adoption of a mass movement approach might prove to be a major catalyst for change in the South African food system" (Drimie and Pereira 2016, 20). The following section sketches out further food sovereignty initiatives at the national level and refers to the prevailing challenges.

Further endeavours and urban issues

Some of the difficulties that the SAFSC faces in connecting different organisations and particularly in involving grassroots voices were drivers of the South African Food Sovereignty Dialogue. In late 2016, the Rosa Luxemburg Stiftung's Southern Africa office initiated a "food sovereignty dialogue" between social movements and NGOs. An expert of this office mentioned that their intention was to identify different social groups first and then to create a dialogue to "develop policy contents, to engage in the field of labour laws in agriculture for example farm workers' rights, and to strengthen those working towards food sovereignty".[23] In contrast to the wide thematic fields of the SAFSC, the focus of the Rosa Luxemburg Stiftung is on farm workers.

The motivation to connect so-called movement brokers and gatekeepers underlines the perceived size of the incipient field of food sovereignty at that time. What unites most of the engaged organisations and campaigns is the fact that they are "middlemen" and thus are not directly involved in the struggles and lived realities on the ground. This problem was also reflected in earlier people-centred land reform attempts, for instance in the context of the NLC (Greenberg 2004).

While food sovereignty in South Africa is a banner under which different NGOs and initiatives gather, this development has to be viewed with caution. Jacobs for instance warns, "In many African countries NGOs outnumber social movements and have become the primary vehicle for peasants and workers to express their aspirations" (2013, 7). His experience highlights that the wide spread of NGOs limits the mobilisation of grassroots' movements. In this regard, La Vía Campesina has also critically reflected upon (intermediary) NGOs' efforts to represent small farmers, which is summed up in the popular saying, "not about us without us" (Borras 2008, 204). An expert from ACB spoke of a disconnect in South Africa: "it is a kind of professional class [e.g. NGOs, universities] and the grassroots struggles. [...] We are struggling to make that connection".[24]

Siholonganye highlights similar problems in the country's post-apartheid land struggles; many of those at the grassroots level are illiterate, and are not familiar with exclusionary economic dynamics globally; in contrast, elitist capture of the debates is evident, which limits the protest spirit (Sihlongonyane 2005, 158). In this way, scattered food sovereignty endeavours in South Africa are not comparable with more politicised groundswell movements like the União Nacional de Camponeses, the small farmers' movement in Mozambique (ibid.). Based on the discussions with experts for this book, it is considered challenging to unite several nascent food sovereignty struggles on different fronts, for instance for land redistribution, broader support through extensions services, or research on the commercialisation of seeds and the spread of GMOs. Whether the new, highly visible, and huge C19 People's Coalition (see South Africa's commercialised agri-food system) may be in the position to unite different struggles and propel food sovereignty politics, and if so, how and to what extent, remains to be seen at the time of writing.

When seeking food sovereignty advocates in South Africa, it became evident that not every group using the food sovereignty rhetoric follows the broader ideals of the concept. Some of the social movements and peasants' associations in Southern Africa which have been affiliated with La Vía Campesina are part of the Southern African Confederation of Agricultural Unions (SACAU). However, SACAU is in favour of large-scale commercialised agriculture including GMOs and has the white-dominated commercial farmers' union AgriSA under its wing, which has been accused of land grabbing (Jacobs 2013, 4). Thus, the position of SACAU is somewhat in contradiction with the origins and politics of La Vía Campesina. Such contradictions require further attention when dealing with food sovereignty in South Africa.

In the context of the specific research interest in food sovereignty engagement in urban areas, it has become evident that recent food sovereignty interventions need a stronger interlinkage between urban and rural. While some organisations, for instance the SAFSC, SPP, and FSC, already refer to the connection between urban and rural agrarian issues and thus seek food sovereignty on both fronts, the urban food sovereignty arena, especially, still seems to be relatively unexplored terrain. Prevailing food sovereignty debates in South Africa overlook a plethora of urban food problems and their interconnection with rural problems. It was only in May 2019 that food sovereignty efforts in urban areas received first publicly visible attention from the SAFSC through a student-led food sovereignty initiative at the University of the Witwatersrand, Johannesburg. A newsletter was produced which highlighted food sovereignty attempts to improve nutrition through urban gardening and a farmers' market with local small-scale farmers (South African Food Sovereignty Campaign 2019, 17 ff.). Moreover, the urban and urban–rural connection could be developed further in the SAFSC's Peoples' Food Sovereignty Act. Thus far, the act remains quite general and highlights the need for "proper spatial planning [...] to ensure the development of a food sovereignty system in rural and urban areas" (South African Human Rights Commission and Friedrich Ebert Stiftung 2018, 14). Beyond that, it calls for access to clean water in urban areas and funding for urban agriculture. In light of the experiences of the urban agriculture initiative in George, which is in the focus of this book, the Peoples' Food Sovereignty Act could be developed further, for instance in demanding the improvement and integration of the informal food sector in urban areas. Specific representation and consideration of issues and alternatives in urban areas and to urban–rural interplay in the agri-food system seem urgently needed in the country's dynamic and prevalent food sovereignty discourse.

While initial steps have been taken for food sovereignty to become a galvanising concept to unite different actors, localised struggles require further attention. Deprived agrarian realities within and on the fringes of cities, including those of marginalised consumers, remain largely absent from the bigger food sovereignty picture and well-articulated discourses. While there have been several cases of food price protests and land occupations in townships, these are not part of a wider insurgence and are often tamed by small-scale interventions such as food kitchens or food gardens, in some cases subsidised by the government. Food sovereignty discourses remain strongly dominated by NGOs; they can be considered invisible for grassroots initiatives. This is illuminated with the case of the KEF initiative in George. In line with the expert interviews for this study, Satgar and Cherry highlight that it is key to "widen, deepen and strengthen local food sovereignty pathways" and to propel "movement building" (2020, 331). This calls for in-depth exploration for food sovereignty efforts on the ground.

Scholars and activists a like keep on pointing out that food sovereignty is not only a rural imperative but also an urban struggle and thus it is key to move beyond primarily engaging with rural populations and to consider expressions across rural and urban (Bowness and Wittman 2020; Jacobs 2013). Fragmentation, weak organisational structure, and the difficulties of linking up with the realities on the ground reflect the role of social movements in land reform debates in the past. A central challenge confronting land movements in South Africa is organisation from below, the relationship between different organisations and movements, and the forms of pressure on the state at different levels (Hall and Ntsebeza 2007, 17). It is thus important for food sovereignty efforts and alliances to learn from localised and grassroots' experiences.

Notions of urban agriculture and introduction of the case study

The work of urban food producers shows many parallels to concrete food sovereignty actions, and it is hence relevant to understand the roots and role of urban food production. In several areas, these small-scale farmers rather seem to operate in niches and food production in the city is often of contested nature. In South Africa, historical developments play a key role here. Cities are still strongly shaped by the history of institutionalised racial segregation. In 1923, the Native (Urban Areas) Act was implemented, which "defined an African's legal ability to be in urban areas as contingent upon employment and empowered White authorities to control non-Whites' access to urban areas" (Lanegran and Lanegran 2001, 673). Permitted black workers were only allowed to reside in segregated settlements at the fringes of the cities. In 1950, the Group Areas Act further allowed white authorities "to remove people living in the 'wrong' areas", although the act was never fully implemented (ibid.). Other measures included apartheid's "influx control" laws implemented in 1948, which prevented permanent settlement of the black population in cities until 1986 (Zhan and Scully 2018, 1022). Most of these laws were softened towards the end of apartheid, and directly afterwards, urbanisation rates increased tremendously. Overall, the demand for this so-called migrant labour power can be traced back to the transformation of largely agrarian societies to industrialised and capitalist societies.

Today, ongoing dynamics of rural–urban migration and growth "from within" are expected to further increase the country's urban population. Growth mainly takes place in marginalised urban areas which feature increasing un(der)-employment, food insecurity, and poverty as illuminated above. Hence, retaining access to land and to some extent preserving the features of an agrarian society can be observed in cities. Considering these urban developments and the previously outlined dynamics of agrarian change from the viewpoint of the critical urban food perspective, urban agriculture presents variegated response to the inequalities enshrined in the corporate agri-food system and neoliberal urban developments.

Crush et al. indicate that, in South Africa "household food production reportedly escalated following the end of apartheid with continued city growth, increasing levels of food inflation and rising unemployment within the formal economy" (2011, 286). Beyond the role of food cultivation as a self-help strategy, the FAO highlights its diversity: "Urban agriculture describes crop and livestock production within cities and towns and surrounding areas. It can involve anything from small vegetable gardens in the backyard to farming activities on community lands by an association or neighbourhood group" (2010, 1). As Maxwell and Zziwa point out, urban agriculture is "as old as African cities themselves" (1992, 13). Given its fragmented nature, often illegal character, and sparse documentation, it can only be estimated that urban farming is practiced by 30–70% of the poor in urban areas in the Global South (FAO 2010, 1). The existing urban agriculture landscape in South Africa is clearly eclectic and emerges from diverse motivations. Urban agriculture typically covers an array of food production and consumption processes, comprising small-scale urban food production, community assisted agriculture, community gardens, school gardens, backyard gardening, farmers' markets, and food collectives. Citizens with highly diverse backgrounds are producing food in cities, ranging from small-scale, peasant-like farming to commercial rooftop farms. In some affluent neighbourhoods, urban agriculture has become an integral part of ethical and conscious consumerism.

An extensive body of literature on urban food production reveals that motivations and conditions of food producers are highly diverse globally (e.g. Clendenning, Dressler, and Richards 2016; McClintock 2011). Inspired by McClintock's typology of urban agriculture, Table 2.1 provides a rough overview of the most common types of food production in cities in South Africa drawing fieldwork in 2016 and 2017. The table presents these types along with their scale of production, functions, management, labour, and market engagement. It includes specific examples from the case study, the KEF initiative in George (see last column). These categories are not static, and many types share similarities or are even combined, as the example of KEF shows in the following sections.

Table 2.1 Most common types of urban agriculture in South Africa including the "Kos en Fynbos" initiative (adapted from McClintock 2014, 150, and own investigation)

Type of urban agriculture	Scale of production	Primary functions and orientation	Management	Labour	Market engagement	Examples from South Africa	"Kos en Fynbos"
Residential	Yards	Food production mostly for own consumption, diversifying diets, recreation, occasional sales of surplus	Individual or household	Self or family	Minimal	In George, residential gardening is common all over the city including low-income and wealthier areas.	Yes: most of the KEF members are growing food in their backyards, some sell the produce informally
Guerrilla	Vacant lots, plants (e.g. seed bombs)	Food production mostly for own consumption, reclaiming space, greening and fighting the concrete jungle	Individual or collective	Individual or collective	Rare	Guerrilla gardeners of Tyisa Nabanye (NGO), Cape Town, growing vegetables on a piece of abandoned military land and are claiming official tenure	Partly: some are gardening on available space on the roadside
Institutional (e.g. schools, prisons, hospitals)	Yard or other vacant space	Food production for public provision in institutions, diversifying diets, education, rehabilitation, greening and fighting the concrete jungle	Institution and/ or volunteers	Institutional members (e.g. patients, prisoners), staff, volunteers	Rare	Blanco Clinic and several creches in George: some have smaller beds others have an extended food garden, mainly managed by the staff with support of volunteers, pupils, parents etc.	Yes: KEF started some of these gardens, runs some of them or supports them

(*Continued*)

Table 2.1 (Continued)

Type of urban agriculture	Scale of production	Primary functions and orientation	Management	Labour	Market engagement	Examples from South Africa	"Kos en Fynbos"
Non-profit/ collective	Vacant lots, parks	Food production, diversifying diets, education, greening and fighting the concrete jungle	Non-profit organisation, society, community garden manager	Staff, volunteers, collective/ community members	Depends	Botanical Garden, George: managed by the Botanical Garden Society, includes a permaculture site which is used for educational purposes; harvest is not sold at the market; Nyanga People's Gardens, Cape Town: community gardens in an informal settlement	Yes: KEF supports the permaculture site of the Botanical Garden and supports a few community gardens
For-profit: small scale at the margins of the city	Large parcels, might include land occupations	Food production incl. livestock raising; income and household food provision, often part of the informal food economy	Individual, some farmers are organised in groups	Self, family, members of the group	Frequent	Sandkraal farm, Thembalethu, George: most of the small-scale farmers are organised in the Thembalethu's Farmers Union; unsecure land access is an issue; Philippi horticultural area, Cape Town	Yes: a few KEF members are engaging in small-scale farming in Thembalethu and sell the produce informally
For-profit/ commercial: in wealthier areas	Large parcels and greenhouses	Food production and alternative food markets, edible landscaping, greening	Business owner and/or manager	Employees	Always	Oranjezicht City Farm, Cape Town: active in an affluent neighbourhood, produce is sold at the Farmers' Market at the Waterfront	No

For deprived working people – particularly those who are underpaid and without (any kind of) social protection (Marcuse 2009, 190) – urban agriculture has special relevance. This relates to the capitalist impulse to keep wages low to subsidise capital and deprived workers' tendency to over-exploit their own labour power to generate more labour output (Shivji 2017, 10). This is a defining feature of working people: they keep expenditures for social reproduction to a minimum, and thus "cheap food" is a key pillar of this so-called system of "super-exploitation" (ibid., 11). Hence, precarious working conditions and low wages contribute to a crisis in social reproduction. It is in this specific context that urban agriculture in South Africa gained momentum as a vital self-help mechanism and livelihood strategy for many, from those who are not (fully) integrated in the labour market to those relying on limited social grants (Crush, Hovorka, and Tevera 2017; Olivier and Heinecken 2017, 746; Paganini, Lemke, and Raimundo 2018, 410). This is also emphasised by increasing land demand in urban areas. A comprehensive survey on land demand in South Africa from 2005/2006 indicates that about one-third of black Africans sought land for agricultural production, and 34% of this demand was from urban areas (Aliber, Reitzes, and Roef 2006). These lived realities often connect with the earlier highlighted dynamic process of semi-proletarianisation, which refers to a combination of land- and labour-based livelihoods (Jacobs 2018, 15). In the context of South Africa, Zhan and Scully highlight that urban wage work has become more insecure, which further increases the importance of the land for food or income – in general as an overall safety net (2018, 1018). The concept of semi-proletarianisation and related observations regarding social production are essential considerations in the critical urban food perspective.

While research from South Africa reminds us that the potential of urban agriculture to fight food insecurity and poverty per se is limited, it can supplement and diversify household food provision and livelihoods (Crush, Hovorka, and Tevera 2017; van Averbeke 2007). Over recent years, research on urban agriculture in Sub-Saharan Africa has focused on its manifold possible benefits in the realms of the social (e.g. community cohesion, social capital, women's empowerment), ecological (e.g. urban greening), health (e.g. dietary diversification) and economic (e.g. income) (e.g. Battersby and Marshak 2013, 448; Engler, Köster, and Siebert 2014, 10–11; Olivier and Heinecken 2017). Following these positive assessments, material and rhetorical support have come from many international organisations, scholars, local and national governments, which have increasingly emphasised the role of urban agriculture in feeding the cities, creating employment, and generating income for the urban poor (Badami and Ramankutty 2015, 8; FAO 2011, x; Zezza and Tasciotti 2010, 266). The City of Cape Town's Urban Agriculture Policy, for instance, reflects these assumptions: "The city believes that urban agriculture can play a pivotal role in poverty alleviation (to improve household food security and nutrition status of people) and economic development (as economic activity it can contribute to job creation and income generation)" (City of Cape Town 2007, 4). However, institution-led work on urban agriculture often remains the main intervention through which hungry and un(der)employed population groups are targeted (Badami and Ramankutty 2015, 9; Zezza and Tasciotti 2010, 266; Webb 2011, 195).

Over the last two decades, several studies have been conducted on urban agriculture and food systems in metropolitan areas in South Africa (e.g. Crush, Hovorka, and Tevera 2017; Olivier and Heinecken 2017; Paganini and Lemke 2020). However, little is known about smaller towns, so-called secondary and intermediate cities (Marais 2016; South African Cities Network - SACN 2017). Even more strikingly, the mobilisation of urban food producers and their political ideas have been barely considered in previous studies. This book, particularly the critical urban food perspective, attempts to shed light on these lacunae.

Critical and differentiated research on institution-led food production in cities around the world indicates that many urban agriculture projects fail as they do not meet the social and economic needs of the participants (e.g. Battersby 2013, 459; Paganini, Lemke, and Raimundo 2018, 411–412; Shillington 2013, 109). In fact, local authorities often have limited means to target the root causes of poverty or to interact closely with the participants. Simultaneously, beyond isolated interventions on project level, local governments tend to undervalue, overlook or even restrict food cultivators at the margins of cities. Food poverty in urban areas in Sub-Saharan Africa has been neglected by the government for a long time and has seldom been investigated intensively by researchers (Battersby 2013, 454). In this regard, Crush and Riley refer to a "rural bias" and explain that food insecurity is still predominantly considered a rural problem in the international as well as the South African development agenda (for instance in the Sustainable Development Goals) (Crush and Riley 2018). However, agrarian and food issues across the rural and urban divide have always been intertwined. As highlighted earlier, deprivation in the countryside leads many people to migrate to the cities. But urban un(der) employment causes further problems, ranging from hunger to high availability of cheap labour power which keeps wages low and limits remittances to the countryside (O'Laughlin et al. 2013, 6). Wider food issues including the affordability and accessibility of food as well as the generally precarious conditions of urban agriculture urgently need to be considered in the urban development and planning agenda (cf. McClintock 2014, 158).

Most urban agriculture takes place on marginalised and unutilised land, for instance on fallow land or small plots between houses or industries or close to roads and railways (Borras 2016, 20; McClintock 2014, 148). Restricted access to land, water and (formal) markets, infertile soil, high and volatile food prices are additional challenges facing urban food producers. On a global scale, the contribution of urban agriculture to total food supply is considered as rather small, helping to feed up to 800 million people (Edelman et al. 2014, 919). In fact, output produced through urban food cultivation and livestock raising is difficult to estimate and most of it probably never enters into official statistics (Borras, Franco, and Suárez 2015, 608).

Some scholars argue that urban agriculture is not to be considered as beneficial per se. For instance, Tornaghi introduces a "critical geography of urban agriculture" rooted in the global north and outlines how food and urban agriculture initiatives could even become a stepping stone for further inequalities: "we need to scrutinize more closely the way these initiatives are becoming, directly or indirectly, new tools or justifications for a new wave of capital accumulation (new green development),

[…] and disinvestments in disadvantaged neighbourhoods (cuts to health and wellbeing services […]" (2014, 553). Clearly, urban agriculture as a locally restricted and sometimes short-term intervention is not able to cure far-reaching neoliberal ills and a troubled agri-food system. At the same time, it was never the aim of many activists and initiatives to critically engage or challenge these inequalities. As outlined in Table 2.1, the various types of urban agriculture have multiple purposes for different groups of people. In the context of community development, McClintock claims that engaging the poor in urban agriculture fills the gaps left in the social safety net by the welfare state, fostering entrepreneurialism and self-help by shifting responsibility to individuals (2014, 155; Pritchard et al. 2016, 705). It is essential to take these messages to the Global South.

However, in the growing body of literature on urban agriculture in the Global South there has been little reflection on the specific role it plays in taking a critical approach to food and land politics. Moreover, urban agriculture can be a means to create and adapt space according to people's needs (i.e. place-making function), demand inclusion, reject governmental interventions or corporate interests (e.g. Ghose and Pettygrove 2014; McClintock 2011). It may embody people's wishes to connect with nature, to see where their food is coming from, to control the production, and even to consume the products of their work and soil. Different experiences of urban food producers show that they reshape urban landscapes, experiment with alternatives to the capitalist organisation of urban life and the industrialised agri-food system. Borras, Franco, and Suárez argue that urban food producers "can transmit ideas of solidarity and struggle against the dominant food system; […] they can amplify public awareness of problems associated with the dominant food system, such as health concerns, the ills of concentration of corporate wealth and power, and so on" (2015, 608–609). Building on this assertion, it can be argued that the political value of these urban food producers can be more substantial than their direct contribution to the food supply. Therefore, food producers might be considered as change agents resisting and confronting the urban ills, or at least revealing them to the wider public. In the long run, so-called local resilience may even engage with and contribute to food system transformation globally (Sage 2014). Indeed, the background and motivations of urban agriculture initiatives, as well as the interaction with other actors in the food system, call for further investigation into the political messages of organised groups.

As briefly outlined in the introduction, the focus of the following chapters is the lived realities of alternative food producers in George. KEF is a diverse group of urban inhabitants engaged in small-scale, ecological food production mainly in backyard gardens, in some cases on fallow farmland and institutional land (i.e. school, hospitals, and crèches). The focus is on the coloured communities of Blanco and Pacaltsdorp as well as the black community in the township of Thembalethu, which are all located at the fringes of the city and thus share an urban–rural character. In George, the legacy of apartheid, particularly in terms of spatial segregation and economic exclusion, can still be felt today. In response to that and in an attempt to create food security, KEF brings different communities closer together and encourages urban agriculture. To better understand and appreciate the work and struggles of the initiative, it is key engage with the specific location, key historical developments, and the contemporary socio-economic situation in George.

George

Approximately 200,000 people were living in George according to the last census in 2011 (Statistics South Africa 2011b). This places George in the category of secondary and intermediate city (SACN 2017, 20). The annual population growth rate of 2.59% (2001–2011) is slightly higher than the national rate (Statistics South Africa 2011b; Western Cape Government Provincial Treasury 2015, 8). Particularly after the removal of apartheid's "influx control" laws in 1986, an ever-increasing number of rural poor labour migrants arrived in George (SACN 2014, 7–8). In addition, many affluent white people have chosen the area for a second home or a retirement spot, adding to the population growth (ibid.). Following these developments, the distribution of population groups in the George Municipality in 2011 was as follows: 50.41% coloured, 28.23% black African, 19.69% white, 0.48% Indian/Asian, and 1.19% other (Statistics South Africa 2011b).

George is located halfway between the metropolitan areas of Cape Town and Port Elizabeth in the south of the country (Figure 2.1). It is part of the Garden Route District in the Western Cape Province. Agricultural land, commercial forests, and national parks surround the city of George. It is situated between the Indian Ocean to the south and the Outeniqua mountains to the north. With its expansive mountain range, rich forests, diverse coastline, lakes and rivers, and fertile land the area is considered a biodiversity hotspot (SACN 2014, 47).

Attracted by these beneficial natural and agricultural conditions, George has been of interest for diverse indigenous tribes and white settlers (mainly of Dutch

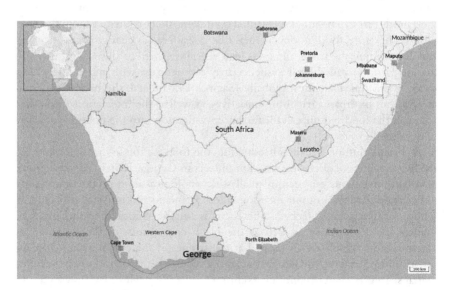

Figure 2.1 Map of South Africa, Western Cape, and George. Prepared by the author (data source: mapz.com 2019, OpenStreetMap (ODbL)).

and British descent) in the colonial and post-colonial eras. It is assumed that the Khoikhoi[25] arrived about 2000 years ago from the north, particularly the area of the modern Botswana. The Khoikhoi were semi-nomadic pastoralists and raised sheep and cattle. They were well known for their honey, which they collected from the fynbos on the slopes of the Outeniqua[26] mountain range. It is said that the San tribe, mainly hunters and gatherers, previously occupied the area and tensions arose between the two tribes (ibid. 4).

In the colonial era, white settlers encroached the land, and the area became well known for its farms, with many of the Khoikhoi eventually becoming farmworkers. Again, there was tension between the indigenous tribes and the new arrivals (ibid. 4). Between 1730 and 1809, the Dutch East India Company allocated loan farms to settlers in the heart of the Outeniqualand. These historical developments show that farming and livestock raising have deep roots in the area. In the past, cultivation mainly focused on wood, livestock, wheat, hop, and barley. Besides the relatively prosperous farmers, the area attracted poor woodcutters. In the late seventeenth century, the Dutch East Indian Company established a woodcutter post, which marks the starting point of commercial woodcutting. All of this contributed to the recognition of the place. George was officially established in 1811 by British settlers. Only two years later the London Missionary Society founded a mission in George which contributed further to George's recognition.

George's roots in farming are still reflected in its contemporary spatial composition. Outside the city centre, the municipality comprises tiny rural agricultural settlements and sparsely populated commercial farming areas (SACN 2014, 96). The rather rural settlements of Herold's Bay, Victoria Bay, Haarlem, Hoekwil, Thembalethu, Touwsranten, Uniondale, and Wilderness are part of the municipality. With the continued growth of the city, previously independent farming villages, for instance Pacaltsdorp and Blanco, became part of George (Siebert and May 2016, 160). These suburbs still have a strong rural character with large swathes of open fields surrounding them. The largest suburb of George is the township Thembalethu, which has its roots in the apartheid era, and similarly features open farmland at the fringes.

The apartheid regime amounted to a policy of white supremacy country-wide. In George, too, apartheid established and perpetuated a "race- and class-based spatial urban order" (Lanegran and Lanegran 2001, 671). As part of this racial segregation, Blanco, Pacaltsdorp, and Thembalethu became distinct settlements. While the old farming villages Blanco and Pacaltsdorp were classified as coloured communities, black people had to reside in the historically informal settlement Thembalethu. Particularly in the early apartheid period, informal settlers were not permitted on the George commonage (today's city centre). However, by the mid-1970s, the municipality of George could no longer ignore that these dwellers were integral to the city's economy; many were working for the municipality or the South African Railways. Thus, the municipality officially began planning a township in 1982 (Lanegran and Lanegran 2001, 679). The city centre itself was declared a "whites only" area. The past racial geography played out in diverse spheres of everyday life including weak access to infrastructure and services for black and coloured communities (SACN 2014, 97–98). At the same time, the majority of the population was systematically excluded from participating meaningfully in the economy and the apartheid regime denied them political rights (Faling 2012, 166).

George's economic developments during the apartheid era still shape the city today. At the end of the 1970s, the establishment of the regional airport in George fuelled the city's role in tourism and in the second homes industry as part of the Garden Route region (Marais 2016, 76). In 1982, the city was identified as one of the industrial development points in the government's Regional Industrial Development Programme which aimed to stimulate economic activities and attract entrepreneurs (SACN 2014, 6). These attempts paved the way for the city to become a major economic player in the region (ibid.).

In 2016, the gross domestic product growth rate of the George Municipality was 1.7% and thus higher than that of the province (1.2%) (Western Cape Government Provincial Treasury 2018, 356). George has become a commercial hub in the Garden Route district and thus plays an important regional service role for the rural hinterlands especially in education, and the provision of health services and consumer goods (SACN 2014, 73–47). It was also in this context that the tertiary sector received more attention. Today, wholesale and retail trade, accommodation, as well as finance, insurance, real estate, and business services contribute more than 70% to the GDPR (Western Cape Government Provincial Treasury 2018, 358); about half of the working population are employed in these sectors (ibid. 364). Directly related to these sectors is a profitable retirement industry and a strong construction sector. In the past, George's economy relied heavily on the primary sector (i.e. agriculture, fishing, and forestry), which contributes only about 4% of the GDPR today (Western Cape Government Provincial Treasury 2018, 359; SACN 2014, 14). Only 10% of the working population were employed in this field in 2016 (Western Cape Government Provincial Treasury 2018, 364). This reflects the country-wide commercialisation of agriculture and the focus on large-scale agriculture, which has seen a reduced demand for agricultural labour since the end of apartheid.

After the overthrow of apartheid, its legacy of wide-ranging racial exclusions called for an urgent socio-political transformation and a democratic turn. In George, scholars have observed that "the political communication seemed to be open and cordial" (SACN 2014, 54). Lanegran and Langran even claim that "the political transformation of George in the 1990s has been remarkable" (2001, 672). While in many other South African cities so-called white flight occurred at the end of apartheid, this was not the case in George (SACN 2014, 54). Nevertheless, local authorities quickly opened up for non-white members and employees (ibid.). In the first democratic local elections in 1995, the ANC and its allies came to power; but many coloured voters shifted their support to the Democratic Alliance in the local authority elections in 2011 (ibid., 97–98). As in many cities in the Western Cape, the Democratic Alliance, which draws its support strongly from coloured and white voters, was victorious in George.

Despite these shifts in local government, the apartheid legacy continues to play a role through spatial and economic exclusions. George shares many similarities with the country as a whole. In the context of rapid growth, it followed the example of other metropolitan areas in urban integration, e.g. "decentralisation of business space, [continued] class-based segregation and gated development" (SACN 2014, 5). However, these attempts turned out to be inappropriate in this intermediate city with its fragmented outskirts. About 15% of the population is living in informal dwellings with poor basic service provision (Statistics South Africa 2011b). In response, governmental development programmes like subsidised low-income housing are intended to improve their conditions (Lanegran and Lanegran 2001, 672). In 2014, about 23,000 persons were on the waiting list for state-subsidised housing (SACN 2014, 98). Given the ever-increasing influx of rural labour migrants, this number has been ever increasing. Spatial restrictions on marginalised regions far from the centre add to these exclusions (Lanegran and Lanegran 2001, 671). These developments urged the city to implement an Integrated Public Transport Plan, which seems to function better than in other cities (SACN 2014, 98).

Marginalised groups continue to be trapped in a vicious cycle of deprivation, which may include unequal access to food, education, health services, family planning, the labour market, or improved housing (Gradin 2013, 187). Food insecurity rates reflect the provincial average. In 2013, about one-quarter of the population was at risk of hunger and about 16% were experiencing hunger; prevalence rates for being at risk of hunger are much higher for non-white population groups (Shisana et al. 2013, 146).

In general, marginalisation is interwoven with inadequate welfare and social services and particularly weak integration in the economy. George's unemployment rates are slightly lower than the provincial average with 15.3% in 2016 (Western Cape Government Provincial Treasury 2018, 367). A large portion of these unemployed people are unskilled and unemployed rural migrants from the Eastern Cape and other African countries (SACN 2014, 4 and 7). In response, a vibrant informal sector has been playing an important role in income and employment creation as well as consumption in George and many other South African cities (SACN 2014, 26 ff.; Skinner and Haysom 2017). However, official numbers do not exist, and estimates vary widely.

As elucidated earlier, economic exclusions and food insecurity cause many households to engage in urban agriculture as a self-help strategy. Data from the 2011 Census (Statistics South Africa 2011a) provide important information on agricultural households at the municipal level, and the South African Community Survey 2016 (Statistics South Africa 2016) shows trends at the provincial level. The South Africa Agricultural Households Community Survey broadly defines an agricultural household as, "[a] household involved in agriculture" (Statistics South Africa 2016, 34). Households were selected if they "produce[d] any kind of food or other agricultural products (e.g. livestock, poultry, crops, food gardening, forestry, fish) whether sold or consumed" (ibid. 1). In the George Municipality, about 8% of the households were engaged in agriculture in 2011 (Statistics South Africa 2011a). Most of these households were active in vegetable production, with smaller numbers in livestock and poultry production (ibid.). The data show that it is mostly people with lower incomes and lower levels of educational attainment that engage in agriculture. Specific reasons as to why they pursue this activity are not covered. But the community survey from 2016 shares insights on the main purposes of agriculture activities in the Western Cape Province; the majority of the households pursue agriculture as an extra source of food (31.8%) or as a leisure activity (26.9%); for one-quarter of the participants it was the main source of food (Statistics South Africa 2016, 17). The community survey moreover indicates different places for these activities. Most of the households use their backyards (80.9%) and about 15% harvest on farmland (ibid.). Although precise definitions of these places (e.g. land size) are not provided, this information suggests many parallels to the KEF initiative and thus helps to understand the urban/peri-urban agricultural landscape in George.

Kos en Fynbos

While food producers favour the rich and fertile soil along the Garden Route, KEF started as a response to the increase in nutritional deficiencies occurring amongst dwellers of poor communities (Siebert and May 2016, 159). The so-called groundswell movement intends to create healthier communities including solidarity and access to nutritious, locally grown food. The members cultivate urban ground in different ways: many grow food in their backyards; some maintain gardens in schools and the hospital; yet others occupy farmland on the fringes of the city. This reflects the information provided in the South African community survey 2016. Most of the members are growing vegetables, salad, and herbs for themselves behind or in front of their houses. Backyard gardens, between 4 and 10 square metres in size, and small door gardens (literally the size of a door), are most common. Only a few people sell their produce so food production is mostly a socially reproductive practice. The initiative was started by a small group of affluent white people (e.g. NGO workers, a doctor, and professors). Today, it mainly consists of coloured members, and a smaller number of black and white members. The majority comes from the deprived working class. Following Shivji's thoughts, the initiative's members can be considered as so-called working people. Their lived realities are in the focus of the book's critical urban food perspective.

Besides everyday practices of food production, the movement organises workshops and regular meetings to exchange knowledge, experiences, and the literal fruits of their labours. In sum, members create solidarity networks, strengthen self-reliance and local production and consumption. As part of the daily farming practices, the movement demands access to land and healthier food, which has caused tensions with the municipality.

Since 2012, the voluntary initiative has spread all over town. Today, KEF is active in Blanco, Pacaltsdorp (incl. Dellville Park, New Dawn Park, and Rosedale), Thembalethu, La Vallia, Rosemoor, Protea Park, Touwsranten, and Kleinkrantz. The latter two are separate villages outside the city area (Greater George area). The two maps below provide an overview of the location. Figure 2.2 shows George and its surroundings including the communities of Touwsranten and Kleinkrantz in the east. The other communities are located in the Greater George area, which is also called the city area (inside the circle on Map 2.2). Figure 2.3 zooms in on the Greater George area and highlights the KEF communities. Protea Park, Rosemoor, and Le Vallia are in the city centre. Blanco, at the north western edge, and Pacaltsdorp,

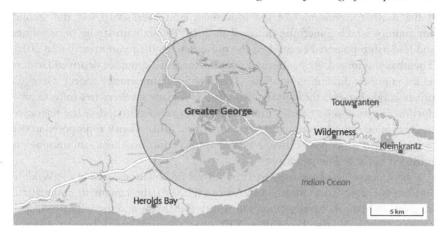

Figure 2.2 Map of Greater George. Prepared by the author (data source: mapz.com 2019, OpenStreetMap (ODbL)).

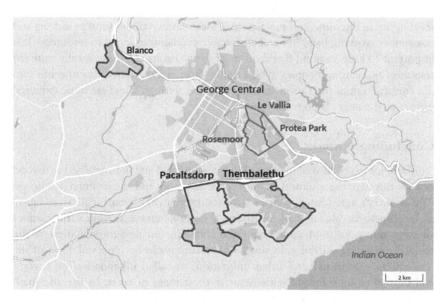

Figure 2.3 Map of the focus areas: Blanco, Pacaltsdorp, and Thembalethu. Prepared by the author (data source: mapz.com 2019, OpenStreetMap (ODbL)).

in the east, were once independent villages outside George. Thembalethu, the large township, is located at the southern outskirts of the city. Most of these communities are considered marginalised given severe shortcomings in adequate service supply and housing as well as unemployment, to name just two of the problems.

In the initial research phase, it became clear that particularly Blanco, Pacaltsdorp, and Thembalethu shape the initiative with their unique character. The other areas joined passively as part of the municipality's urban gardening interventions. Blanco is the mother community of the movement, and Pacaltsdorp was the second community which joined the initiative in 2014. People with strong personalities and leadership potential established the movement in their communities. In 2015, Thembalethu joined – a committed coordinator and a group of deprived farmers added another dimension to KEF. These three communities reflect George's urban–rural character through their location and many dwellers' ties to food production as part of their family's roots. These characteristics underline the relevance of combining critical agrarian studies with critical urban theory as proposed in the critical urban food perspective. Detailed information on these communities is provided in the following chapter.

The initiative has been supported by several organisations, mainly the Wildlife and Environment Association South Africa (WESSA), the Landmark Foundation, the Nelson Mandela Metropolitan University (NMMU), and the local newspaper *George Herald*. At the beginning, KEF was also supported by the George Municipality, particularly the Local Economic Development Unit. However, this cooperation became rather weak in 2016.

While the term food sovereignty has not been used explicitly by most of the KEF members themselves, KEF presents many entry points and linkages to food sovereignty in discourse and practice (e.g. localisation, valuing food producers and consumers, working hand in hand with nature, reclaiming access to resources). It is important to keep in mind that food sovereignty has its origins in locally initiated responses of farming peoples, whose multi-faceted experiences breathe life into the construction of food sovereignty in different settings. These are to be explored further in the next chapters.

Concluding remarks

South Africa has been well integrated in the globalised and industrialised agri-food system through the political economy of its food and agricultural landscape. The country's agricultural policies have followed a neoliberal ideology to satisfy and compete in global markets. Today, a comprehensive commercialisation of all steps along the food value chain and simultaneous de-agrarianisation of the country including a decline in small-scale farming can be observed. One of the consequences remains rural–urban migration, so-called urbanisation of poverty, and ever-growing informal settlements at the fringes of cities. In the aftermath of apartheid, land reform has been a key element in reconciliation policies, but it has only benefited a few and the government has been unable to rectify land inequality in both rural and urban areas. Overall, economic growth is far from being inclusive, which plays out in soaring poverty and unemployment rates. Many people are engaged in multiple (informal) income activities or rely on social grants. These realities at the margins fuel further challenges such as difficulties in sufficient access to healthy and affordable food, leading to partial self-provision, for instance in growing food. Official land access remains another issue. This ties in with the outlined theoretical considerations as part of the critical urban food perspective in the introduction of this book. It is this specific context, which urgently requires further attention in the wide food sovereignty landscape globally. On the whole, the country's agri-food system shows many similarities to other middle-income emerging economies in the Global South e.g. Brazil, India, and Indonesia (Chappell 2018; Li 2015; Purushothaman and Patil 2019).

It can be argued that in South Africa initial steps have been taken for food sovereignty to become a galvanising concept to unite different social groups in the agri-food system. In this context, it is key to consider the work of several land reform organisations and related difficulties. A key challenge of land movements in the past was organisation from below and solid alliances. Hence, the NLC, the LPM, and ALARM are clear examples of endeavours that did not flourish. It is thus important for food sovereignty efforts and alliances to learn from these experiences. Moreover, two food sovereignty campaigns were introduced and NGOs which have integrated food sovereignty prominently into their work. While further organisations are active in the outlined fields (e.g. land and agrarian reform, support of farmworkers), other NGOs and community-based organisations do not necessarily operate under the food sovereignty banner or have close linkages to existing campaigns or to internationally recognised food sovereignty movements. Food sovereignty in discourses remains strongly dominated by NGOs; they can be considered incipient and even invisible for grassroots initiatives. For instance, the SAFSC has only been slowly forming a national exchange platform since February 2015. This kind of fragmentation and the difficulties of linking up with the realities on the ground reflect the role of social movements in land reform debates in the past. In sum, food sovereignty as a wider social movement response to the unequal agrarian structure and food system remains incipient. Food sovereignty discourses do not seem to consider deprived agrarian realities within and on the fringes of cities. To date, food sovereignty initiatives have not taken root deeply in urban and rural areas. However, recent work of the Peoples' Food Sovereignty Act might benefit from experiences like those of the urban agriculture initiative KEF. In fact, voices from below are often not directly on the radar of the civil society organisations described here, or part of their campaigns.

The third part of this chapter illuminates the role of food production in South African cities, which often remains an undervalued self-help strategy and is challenged by difficulties in land, water, and seed access. However, urban agriculture could also be considered a response "from below" by the neglected – those who are not fully integrated in the labour market and those who are struggling to meet their dietary needs. In South African cities, vital food production on empty plots, gardening on the roadside, food cultivation in solidarity projects, livestock raising, and farming at the urban outskirts show that many citizens are more than passive dwellers. These lived realities of urban food producers point to the failures of the country's agri-food system. Against this background, the KEF initiative and particularly the socio-economic background of the city of George are introduced. The medium-sized city has deep roots in farming and still comprises village-like settlements at the outskirts. However, the apartheid legacy is still mirrored in its spatial composition and broader economic exclusions. The initiative critically engages with these conditions. Members of the initiative are working jointly towards better nutrition, land access, and solidarity in and beyond their communities.

Notes

1 Net importer implies that import rates have been higher than export rates for many agricultural products (mainly fertiliser and machinery) and several foods products, for instance wheat, rice, palm oil, corn, seeds, poultry, and dairy products (see Greenberg 2017; Pritchard et al. 2016, 704).

2 From 1652 onwards, the colonial era comprises several periods of Dutch rule and British rule.

3 Homelands are also referred to as Bantustans and reserves.

4 Over the last decades, several acts and regulations were introduced that maintain unequal seed and thus agri-food systems (e.g. the Plant Breeder's Rights Act in 2018). The African Centre for Biodiversity (ACB) has done intensive research in this regard.

5 Convenience food points to a high content of "sugars and sweeteners, salt, fats, and refined grains; [and] the ubiquitous availability of these products; and very large advertising budgets to market these products" (Scrinis 2015, 137).

6 Four main strategies were introduced to target food security at the national level, which are in line with the right to food. Those are the Integrated Food Security Strategy (2002), several general targets within the National Development Plan (2011), the Fetsa Tlala (End Hunger) programme (successor of the Zero Hunger Programme, 2009), and the National Policy on Food Security and Nutrition (2013). The first three interventions focused on skills, food transfers, school feeding schemes and cash transfers/social grants, and training in nutrition awareness. While these interventions failed to target the systemic roots of the problem, their implementation was also complicated by mismanagement of funds and corruption. The still ongoing National Policy on Food Security and Nutrition seeks to overcome these shortcomings, and aims to integrate smallholders more efficiently into the market.

7 While the informal sector is difficult to define it is important to go beyond a simple counter-narrative of the formal. Certainly, the informal is closely linked to so-called cracks in the system. However, a simplistic division between formal and informal does not necessarily help and is often not feasible as fine lines often connect the two. While informality in its diverse nuances (e.g. in markets, housing, service supply, food production, land access, livelihoods etc.) persists it is important to engage in graded changes instead of abruptly eradicating these structures.

8 The C19 People's Coalition is a civil society collective committed social justice and democratic principles. The focus is on those who are most vulnerable to the crisis. The collective comprises community and non-governmental organisations, social movements as well as trade unions. It was initiated in March 2020.

9 Stiftung is the German word for Foundation.

10 Interview on March 23, 2017.

11 Interview on March 22, 2017.

12 Interview on March 23, 2017.

13 Interview on September 7, 2016.

14 Interview on September 7, 2016.

15 Interview on March 22, 2017.

16 Interview on March 23, 2017.

17 Interview on September 7, 2016.

18 Interview on March 22, 2017.

19 Interview on March 22, 2017.

20 Interview on March 23, 2017.

21 Interview on March 22, 2017.

22 Interview on March 23, 2017.

23 Interview on March 22, 2017.

24 Interview on March 23, 2017.

25 Khoikhoi translates as "the real people" or "men of men", which relates to the pride of their culture.

26 The name was introduced by the Khoikhoi and means "men who carry honey".

References

African Centre for Biosafety 2009. Biotechnology, seed and agrochemicals: global and South African industry structure and trends. Melville. https://www.acbio.org.za/sites/default/files/2015/02/Biotech-Booklet-11.pdf, Accessed on 20/04/21.

Aliber, M.; Hall, R. 2012. Support for smallholder farmers in South Africa: Challenges of scale and strategy. *Development Southern Africa* 29, No. 4, 548–562. DOI: 10.1080/0376835X.2012.715441.

Aliber, M.; Mdoda, L. 2015. The direct and indirect economic contribution of small-scale black agriculture in South Africa. *Agrekon* 54, No. 2, 18–37. DOI: 10.1080/03031853.2015.1065187.

Aliber, M., Reitzes, M., & Roefs, M., 2006. Assessing the alignment of South Africa's land reform policy to people's aspirations and expectations: A policy-oriented report based on a survey in three provinces. Working paper, Human Sciences Research Council, Pretoria.

Badami, M.G.; Ramankutty, N. 2015. Urban agriculture and food security. A critique based on an assessment of urban land constraints. *Global Food Security* 4, 8–15. DOI: 10.1016/j.gfs.2014.10.003.

Baletti, B.; Johnson, T.; Wolford, W.M. 2008. "Late mobilization". Transnational peasant networks and grassroots organizing in Brazil and South Africa. *Journal of Agrarian Change* 8, No. 2–3, 290–314. DOI: 10.1111/j.1471-0366.2008.00171.x.

Battersby, J. 2013. Hungry cities. A critical review of urban food security research in Sub-Saharan African Cities. *Geography Compass* 7, No. 7, 452–463.

Battersby, J.; Marshak, M. 2013. Growing communities. Integrating the social and economic benefits of Urban Agriculture in Cape Town. *Urban Forum* 24, 447–461. DOI: 10.1007/s12132-013-9193-1.

Bernstein, H. 2011. 'Farewells to the peasantry?' and its relevance to recent South African debates. *Transformation* 75, No. 1, 44–52.

Bernstein, H. 2013. Commercial agriculture in South Africa since 1994: "Natural, simply capitalism". *Journal of Agrarian Change* 13, No. 1, 23–46. DOI. 10.1111/joac.12011.

Borras, S.M. 2008. Reply: solidarity. Re-examining the "agrarian movement-NGO" solidarity relations discourse. *Dialect Anthropol* 32, No. 3, 203–209. DOI: 10.1007/s10624-008-9068-3.

Borras, S.M. 2016. *Land politics, agrarian movements and scholar-activism*. Inaugural Lecture. International Institute of Social Studies, 2016.

Borras, S.M.; Franco, J.C.; Suárez, S.M. 2015. Land and food sovereignty. *Third World Quarterly* 36, No. 3, 600–617. DOI: 10.1080/01436597.2015.1029225.

Bowness, E. and Wittman, H. 2020: Bringing the city to the country? Responsibility, privilege and urban agrarianism in Metro Vancouver, *The Journal of Peasant Studies*, DOI: 10.1080/03066150.2020.1803842.

Bundy, C. 1972. The emergence and decline of a South African peasantry. *African Affairs* 71, No. 285, 369–388.

Bundy, C. 1979. *The Emergence and Decline of a South African Peasantry*. Berkeley and Los Angeles: University of California Press.

Chappell, M.J. 2018. *Beginning to end hunger: food and the environment in Belo Horizonte, Brazil, and beyond*. Oakland: University of California Press.

Cherry, J. 2016. Taking back power in a brutal food system: Food Sovereignty in South Africa. Master thesis. University of the Witwatersrand, Johannesburg. Department of Development Studies.

City of Cape Town 2007. Urban Agricultural Policy for the City of Cape Town. http://www.ruaf.org/sites/default/files/Urban%20agricultural%20policy%20for%20the%20city%20of%20Cape%20Town.pdf, Accessed on 10/02/21.

Claasen, N.; van der Hoeven, M.; Covic, N. 2016. Food environments, health and nutrition in South Africa. Mapping the research and policy terrain. *PLAAS Working Paper* 34, No. Cape Town.

Clendenning, J.; Dressler, W.H.; Richards, C. 2016. Food justice or food sovereignty? Understanding the rise of urban food movements in the USA. *Agric Hum Values* 33, No. 1, 165–177. DOI: 10.1007/s10460-015-9625-8.

Cousins, B. 2010. What is a "smallholder"? Class-analytic perspectives on small-scale farming and agrarian reform in South Africa. *PLAAS Working Paper Series, No.* 16. http:// repository.uwc.ac.za/xmlui/bitstream/handle/10566/4468/wp_16_what_is_a_small-holder_2009.pdf?sequence=1&isAllowed=y, Accessed on 20/04/16.

Cousins, B.; Dubb, A.; Hornby, D.; Mtero, F. 2018. Social reproduction of "classes of labour" in the rural areas of South Africa. Contradictions and contestations. *The Journal of Peasant Studies* 45, No. 5–6, 1060–1085. DOI: 10.1080/03066150.2018.1482876.

Crush, J. 2016. Hungry Cities of the Global South. *Hungry Cities Partnership - Discussion Papers* 1. https://hungrycities.net/wp-content/uploads/2016/06/Hungry-Cities-Final-Discussion-Paper-No-1.pdf, Accessed on 12/02/2021.

Crush J, Hovorka A, Tevera D. 2011. Food security in Southern African cities: The place of urban agriculture. *Progress in Development Studies* 11, No. 4, 285–305. DOI: 10.1177/146499341001100402

Crush, J.; Hovorka, A.; Tevera, D. 2017. Farming in the city. The broken promise of urban agriculture. In Frayne, B., Crush, J., McCordic, C. (Eds.): *Food and Nutrition Security in Southern African Cities*. New York: Routledge, 2018, 101–117.

Crush, J.; Riley, L. 2018. Rural bias and urban food security. In Battersby, J., Watson, V. (Eds.): *Urban Food Systems Governance and Poverty in African Cities*. New York: Routledge, 2018. | Series: Routledge studies in food, society and the environment: Routledge, 42–55.

Drimie, S.; Pereira, L. 2016. Advances in Food Security and Sustainability in South Africa. In Barling, D. (Ed.): *Advances in food security and sustainability*. Amsterdam: Academic Press (Advances in Food Security and Sustainability Volume 1, 1), 1–31.

Drysdale, R.E.; Bob, U.; Moshabela, M. 2021. Socio-economic Determinants of Increasing Household Food Insecurity during and after a Drought in the District of iLembe, South Africa. *Ecology of Food and Nutrition* 60, No. 1, 25–43, DOI: 10.1080/03670244.2020.1783663.

Du Toit, A. 2018. Without the blanket of the land. Agrarian change and biopolitics in post–Apartheid South Africa. *The Journal of Peasant Studies* 45, No. 5–6, 1086–1107. DOI: 10.1080/03066150.2018.1518320.

Du Toit, A.; Neves, D. 2014. The government of poverty and the arts of survival. Mobile and recombinant strategies at the margins of the South African economy. *The Journal of Peasant Studies* 41, No. 5, 833–853. DOI: 10.1080/03066150.2014.894910.

Edelman, M.; Weis, T.; Baviskar, A.; Borras, S.M.; Holt-Giménez, E.; Kandiyoti, D.; Wolford, W. 2014. Introduction: Critical Perspectives on Food Sovereignty. *The Journal of Peasant Studies* 41, No. 6, 911–931. DOI: 10.1080/03066150.2014.963568.

Engler, S.; Köster, J.; Siebert, A. 2014. Farmers Food Insecurity Monitoring Identifying Situations of Food Insecurity and Famine. *IFHV Working Paper* 4, No. 3.

Faber, M.; Drimie, S. 2016. Rising food prices and household food security. *South African Journal of Clinical Nutrition* 29, No. 2, 53–54. DOI: 10.1080/16070658.2016.1216358.

Faling, W. 2012. A spatial planning perspective on climate change, asset adaptation and food security: the case of two South African cities. In Frayne, B., Moser, C., Ziervogel, G. (Eds.): Climate Change, Assets and Food Security in Southern African Cities. Oxon: Earthscan, 163–185.

FAO 2010. Fighting Poverty and Hunger. What Role for Urban Agriculture? FAO. http:// www.fao.org/docrep/012/al377e/al377e00.pdf, Accessed on 02/04/21.

FAO 2011. The Place of Urban and Peri-Urban Agriculture (UPA) in National Food Security Programmes. FAO. http://www.fao.org/docrep/014/i2177e/i2177e00.pdf, updated on 2011, Accessed on 10/02/21.

Food Sovereignty Campaign 2018. Right to Agrarian Reform for Food Sovereignty Campaign: April 17 Statement. https://viacampesina.org/en/right-to-agrarian-reform-for-food-sovereigntycampaign-april-17-statement/, Accessed on 18/10/20.

Ghose, R.; Pettygrove, M. 2014. Urban Community Gardens as Spaces of Citizenship. *Antipode* 46, No. 4, 1092–1112. DOI: 10.1111/anti.12077.

Gradin, C. 2013. Race, poverty and deprivation in South Africa. *Journal of African Economies* 22, No. 2, 187–238.

Greenberg, S. 2004. The landless people's movement and the failure of post-apartheid land reform. *A case study for the UKZN project entitled: Globalisation, Marginalisation*. University of Kwazula Natal. http://ccs.ukzn.ac.za/files/Greenberg%20LPM%20RR.pdf, Accessed on 10/01/2021.

Greenberg, S. 2010. *Contesting the food system in South Africa: Issues and opportunities*. Cape Town: Institute for Poverty, Land and Agrarian Studies, School of Government, University of the Western Cape.

Greenberg, S. 2013. The disjunctures of land and agricultural reform in South Africa: Implications for the agri-food system. August 2013. Bellville: PLAAS, UWC (26).

Greenberg, S. 2014. Agrarian reform and South Africa's agro-food system. *The Journal of Peasant Studies* 42, No. 5, 957–979. DOI: 10.1080/03066150.2014.993620.

Greenberg, S. 2017. Corporate power in the agro-food system and the consumer food environment in South Africa. *The Journal of Peasant Studies* 44, No. 2, 467–496. DOI: 10.1080/03066150.2016.1259223.

Hall, R.; Cliffe, L. 2009. Introduction. In Hall, R. (Ed.): *Another Countryside? Policy Options for Land and Agrarian Reform in South Africa*. Cape Town: Institute for Poverty, Land and Agrarian Studies, School of Government, University of the Western Cape, 1–23.

Hall, R.; Ntsebeza, L. 2007. Introduction. In Ntsebeza, L., Hall, R. (Eds.): *The Land Question in South Africa. The challenge of transformation and redistribution*. Cape Town: HSRC Press, 1–24.

Hart, G. 2013. *Rethinking the South African Crisis: Nationalism, Populism, Hegemony*. Athens: University of Georgia Press.

Hart, G. 2019. From authoritarian to left populism? Reframing debates. *South Atlantic Quarterly* 118, No. 2, 307–323. DOI: 10.1215/00382876-7381158.

Haysom, G. 2016. Alternative food networks and food insecurity in South Africa. *PLAAS Working Paper* 33, Cape Town.

Holt-Giménez, E.; Shattuck, A. 2011. Food crises, food regimes and food movements: rumblings of reform or tides of transformation? *The Journal of Peasant Studies* 38, No. 1, 109–144. DOI: 10.1080/03066150.2010.538578.

Hornby, D.; Royston, L.; Kingwill, R.; Cousins, B. 2017. Introduction: Tenure Practices, Concepts and Theories in South Africa. In Hornby, D., Kingwill, R., Royston, L., Cousins, B. (Eds.): *Untitled. Securing Land Tenure in Urban and Rural South Africa*. Pietermaritzburg: University of KwaZulu-Natal Press, 1–43.

Jacobs, R. 2013. The Radicalisation of the Struggles of the Food Sovereignty Movement in Africa. *La Via Campesina's Open Book: Celebrating 20 Years of Struggle and Hope*. https://viacampesina.org/en/wp-content/uploads/sites/2/2013/05/EN-11.pdf, Accessed on 12/12/20.

Jacobs, R. 2018. An urban proletariat with peasant characteristics: land occupations and livestock raising in South Africa, *The Journal of Peasant Studies* 45, No. 5–6, 884–903. DOI: 10.1080/03066150.2017.1312354.

Jara, M.; Hall, R. 2009. What are the political parameters? In Hall, R. (Ed.): *Another Countryside? Policy Options for Land and Agrarian Reform in South Africa*. Cape Town: Institute for Poverty, Land and Agrarian Studies, School of Government, University of the Western Cape, 207–229.

Kepe, T.; Hall, R. 2018. Land redistribution in South Africa. Towards decolonisation or recolonisation? *Politikon* 45, No. 1, 128–137. DOI: 10.1080/02589346.2018.1418218.

Kirsten, J.F.; van Zyl, J.; van Rooyen, J. 1994. South African agriculture in the 1980s. *South African Journal of Economic History* 9, No. 2, 19–48. DOI: 10.1080/20780389.1994.10417230.

Lahiff, E. 2009. With what land rights? Tenure arrangements and support. In Hall, R. (Ed.): *Another Countryside? Policy Options for Land and Agrarian Reform in South Africa*. Cape Town: Institute for Poverty, Land and Agrarian Studies, School of Government, University of the Western Cape.

Lanegran, K.; Lanegran, D. 2001. South Africa's national housing subsidy program and Apartheid's urban legacy. *Urban Geography* 22, No. 7, 671–687. DOI: 10.2747/0272-3638.22.7.671.

Levin, R.; Weiner, D. 1996. The politics of land reform in South Africa after apartheid. Perspectives, problems, prospects. *The Journal of Peasant Studies* 23, No. 2–3, 93–119. DOI: 10.1080/03066159608438609.

Li, T.M. 2015. Can there be food sovereignty here? *The Journal of Peasant Studies* 42, No. 1, 205–211. DOI: 10.1080/03066150.2014.938058.

Magdoff, F.; Foster, J. B.; Buttel, F. H. 2000. *Hungry for Profit: The Agribusiness Threat to Farmers, Food, and the Environment*. New York: Monthly Review Press.

Marais, L. 2016. Local economic development beyond the centre. Reflections on South Africa's secondary cities. *Local Economy* 31, No. 1–2, 68–82. DOI: 10.1177/0269094215614265.

Marcuse, P. 2009. From critical urban theory to the right to the city. *City: Analysis of Urban Trends, Culture, Theory, Policy, Action* 13, No. 2–3, 185–196. DOI: 10.1080/13604810902982177.

Maxwell, D.; Zzizwa, S. 1992. *Urban Agriculture in Africa: the case of Kampala*. Nairobi: ACTS Press.

McClintock, N. 2011. From industrial garden to food desert. Demarcated devaluation in the flatlands of Oakland, California. In Alkon, A. H., Agyeman, J. (Eds.): *Cultivating Food Justice, Race, Class, and Sustainability*. Cambridge: MIT Press, 89–120.

McClintock, N. 2014. Radical, reformist, and garden-variety neoliberal: Coming to terms with urban agriculture's contradictions. *Local Environment* 19, No. 2, 147–171. DOI: 10.1080/13549839.2012.752797.

McDonald, D. A. 2008. *World city syndrome: neoliberalism and inequality in Cape Town*. New York: Routledge.

Mitlin, D. 2018. Beyond contention. Urban social movements and their multiple approaches to secure transformation. *Environment and Urbanization* 30, No. 2, 557–574. DOI: 10.1177/0956247818791012.

Mngxitama, A. 2006. The Taming of Land Resistance. *Journal of Asian and African Studies* 41, No. 1–2, 39–69. DOI: 10.1177/0021909606061747.

Ngcoya, M.; Kumarakulasingam, N. 2017. The Lived Experience of Food Sovereignty. Gender, Indigenous Crops and Small-Scale Farming in Mtubatuba, South Africa. *J Agrar Change* 17, No. 3, 80–496. DOI: 10.1111/joac.12170.

Ntsebeza, L. 2013. South Africa's countryside: prospects for change from below. In Hendricks, F., Ntsebeza, L., Helliker, K. (Eds.): *The Promise of Land. Undoing a Century of Dispossession in South Africa*. Auckland Park: Jacana Media, 130–157.

O'Laughlin, B.; Bernstein, H.; Cousins, B.; Peters, P.E. 2013. Introduction: Agrarian Change, Rural Poverty and Land Reform in South Africa since 1994. *Journal of Agrarian Change* 13, No. 1, 1–15. DOI: 10.1111/joac.12010.

Olivier, D.W.; Heinecken, L. 2017. Beyond food security. Women's experiences of urban agriculture in Cape Town. *Agric Hum Values* 34, No. 3, 743–755. DOI: 10.1007/s10460-017-9773-0.

Paganini, N., Lemke, S. 2020. "There is food we deserve, and there is food we do not deserve". Food injustice, place and power in urban agriculture in Cape Town and Maputo. *Local Environment* 25, No. 11–12, 1000–1020, DOI: 10.1080/13549839.2020.1853081.

Paganini, N.; Lemke, S.; Raimundo, I. 2018. The potential of urban agriculture towards a more sustainable urban food system in food-insecure neighbourhoods in Cape Town and Maputo. Economic and environmental sustainability and local policy instruments. *Economia Agro-Alimentare* 20, 399–421. DOI: 10.3280/ECAG2018-003008.

Peyton, S.; Moseley, W.; Battersby, J. 2015. Implications of supermarket expansion on urban food security in Cape Town, South Africa. *African Geographical Review* 34, No. 1, 36–54. DOI: 10.1080/19376812.2014.1003307.

Pritchard, B.; Dixon, J.; Hull, E.; Choithani, C. 2016. "Stepping back and moving in". The role of the state in the contemporary food regime. *The Journal of Peasant Studies* 43, No. 3, 693–710. DOI: 10.1080/03066150.2015.1136621.

Purushothaman, S. and Patil, S. 2019. *Agrarian Change and Urbanization in Southern India*. City and the Peasant. Singapore: Springer. DOI: 10.1007/978-981-10-8336-5.

SACN 2014. George Land of Milk and Honey? *SACN Research Report*, University of the Free State. http://www.sacities.net/wp-content/uploads/2015/10/George-report-final-author-tc-3.pdf, Accessed on 22/04/2021.

SACN 2017. Spatial Transformation: Are Intermediate Cities Different? Johannesburg: South African Cities Network. https://www.sacities.net/wp-content/uploads/2021/03/SACN-Secondary-Cities-2017.pdf, Accessed on 23/04/21.

Sage, C. 2014. The transition movement and food sovereignty: From local resilience to global engagement in food system transformation. *Journal of Consumer Culture* 14, No. 2, 254–275. DOI: 10.1177/1469540514526281.

Sanders, D.; Joubert, L.; Greenberg, S.; Hutton, B.; Bautista Hernández, F.A.; Díaz Rojas, I.C. et al. 2014. At the bottom of the food chain: small operators versus multinationals in the food systems of Brazil, Mexico and South Africa. https://foodsecurity.ac.za/wp-content/uploads/2018/04/FINAL-REPORT-MNCs-8-August-2016-SP2.pdf, Accessed on 11/02/21.

Satgar, V.; Cherry, J. 2020. Climate and food inequality: The South African Food Sovereignty Campaign response. *Globalizations* 17, No. 2, 317–337, DOI: 10.1080/14747731.2019.1652467.

Schiavoni, C.M. 2017. The contested terrain of food sovereignty construction. Toward a historical, relational and interactive approach. *The Journal of Peasant Studies* 44, No. 1, 1–32. 10.1080/03066150.2016.1234455.

Scrinis, G. 2015. Big Food Corporations and the Nutritional Marketing and Regulation of Processed Foods. *Canadian Food Studies* 2, 136–145. DOI: 10.15353/cfs-rcea.v2i2.113.

Scully, B.; Britwum, A.O. 2019. Labour Reserves and Surplus Populations. Northern Ghana and the Eastern Cape of South Africa. *Journal of Agrarian Change* 19, No. 3, 407–426. DOI: 10.1111/joac.12309.

Seekings, J.; Nattrass, N. 2006. *Class, Race, and Inequality in South Africa*. Scottsville: University of KwaZulu-Natal Press.

Shillington, L.J. 2013. Right to Food, Right to the City. Household Urban Agriculture, and Socionatural Metabolism in Managua, Nicaragua. *Geography, Planning and Environment* 44, 103–111. DOI: 10.1016/j.geoforum.2012.02.006.

Shisana, O.; Labadarios, D.; Rehle, T.; Simbayi, L.; Zuma, K.; Dhansay, A. et al. 2013. *South African National Health and Nutrition Examination Survey (SANHANES-1)*. Cape Town: HSRC Press.

Shivji, I.G. 2017. The Concept of "Working People". *Agrarian South: Journal of Political Economy* 6, No. 1, 1–13. DOI: 10.1177/2277976017721318.

Siebert, A. 2020. Transforming urban food systems in South Africa: unfolding food sovereignty in the city, The Journal of Peasant Studies 47, 2, 401–419, DOI: 10.1080/03066150.2018.1543275.

Siebert, A.; May, J. 2016. Urbane Landwirtschaft und das Recht auf Stadt. Theoretische Reflektion und ein Praxisbeispiel aus George, Südafrika. In Engler, S., Stengel, O., Bommert, W. (Eds.): *Regional, innovativ und gesund. Nachhaltige Ernährung als Teil der Großen Transformation*. Göttingen: Vandenhoeck & Ruprecht, 153–168.

Sihlongonyane, M.F. 2005. Land Occupations in South Africa. In Moyo, S., Yeros, P. (Eds.): *Reclaiming the Land. The Resurgence of Rural Movements in Africa, Asia and Latin America*. London: Zed Books.

Skinner, C.; Haysom, G. 2017. The informal sector's role in food security. A missing link in policy debate. *Hungry Cities Partnership - Discussion Papers, No. 6*. http://hungrycities.net/wp-content/uploads/2017/03/HCP6.pdf, Accessed on 11/01/21.

South African Food Sovereignty Campaign 2015a. About Campaign. http://www.safsc.org.za/our-story/, Accessed on 10/02/21.

South African Food Sovereignty Campaign 2015b. Report. Johannesburg. https://safsc.org.za/wp-content/uploads/2014/03/FS-Assembly-Report.pdf, Accessed on 10/02/21.

South African Food Sovereignty Campaign 2019. Newsletter #13. May 2019, 13. https://gallery.mailchimp.com/6eb374fe9b580101982b7b47c/files/d15adfa6-0a21-49d6-9933-fa71acddbb84/SAFSC_newsletter_13.pdf, Accessed on 28/03/21.

South African Food Sovereignty Campaign; Friedrich Ebert Stiftung 2018. Peoples' Food Sovereignty Act. https://www.fes-southafrica.org/fileadmin/user_upoad/FS_Act_no.1_of_2018_Short.pdf, Accessed on 17/01/20.

South African Human Rights Commission; Friedrich Ebert Stiftung 2018. Peoples' Food Sovereignty Act. No. 1 of 2018. https://www.fessouthafrica.org/fileadmin/user_upload/FS_Act_no.1_of_2018_Short.pdf, Accessed on 19/03/21.

Statistics South Africa 2011a. Agricultural Households Western Cape Province. Excel File. http://www.statssa.gov.za/?page_id=993&id=george-municipality, Accessed on 10/12/20.

Statistics South Africa 2011b. Census 2011. Population Dynamics in South Africa. Report No. 03-01-67. Pretoria. http://www.statssa.gov.za/publications/Report-03-01-67/Report-03-01-672011.pdf, Accessed on 10/12/20.

Statistics South Africa 2016. Community Survey 2016 Agricultural Households. Report No. 03-01-05. http://www.statssa.gov.za/publications/03-01-05/03-01-052016.pdf, Accessed on 14/02/21.

Statistics South Africa 2020. Quarterly Labour Force Survey. Quarter 4: 2019. Pretoria. http://www.statssa.gov.za/publications/P0211/P02114thQuarter2019.pdf, Accessed on 18/04/2021.

Termeer, C.J.A.M.; Drimie, S.; Ingram, J.; Pereira, L.; Whittingham, M.J. 2018. A diagnostic framework for food system governance arrangements: The case of South Africa. *NJAS - Wageningen Journal of Life Sciences* 84, 85–93, DOI: 10.1016/j.njas.2017.08.001.

Tornaghi, C. 2014. Critical geography of urban agriculture. *Progress in Human Geography* 38, No. 4, 551–567. DOI: 10.1177/0309132513512542.

van Averbeke, W. 2007. Urban farming in the informal settlements of Atteridgeville, Pretoria, South Africa. *Water SA* 33, No. 3, 337–342. DOI: 10.4314/wsa.v33i3.49112.

Webb, N.L. 2011. When is enough, enough? Advocacy, evidence and criticism in the field of urban agriculture in South Africa. *Development Southern Africa* 28, No. 2, 195–208. DOI: 10.1080/0376835X.2011.570067.

Wegerif, M.C.A. 2020. "Informal" food traders and food security: experiences from the Covid-19 response in South Africa. *Food Sec.* 12, 797–800, DOI: 10.1007/s12571-020-01078-z.

Western Cape Government Provincial Treasury 2015. Socio-economic Profile George Municipality. Working Paper. Cape Town. www.westerncape.gov.za/assets/departments/treasury/Documents/Socio-economic-profiles/2016/municipality/Eden-District/wc044_george_2015_sep-lg_profile.pdf, Accessed on 06/02/21.

Western Cape Government Provincial Treasury 2018. Municipal economic review and outlook 2018. https://www.westerncape.gov.za/assets/departments/treasury/Documents/Research-and-Report/2018/2018_mero_revised.pdf, Accessed on 25/02/21.

Williams, M.; Satgar, V. 2020. Transitional compass: Anti-capitalist pathways in the interstitial spaces of capitalism, *Globalizations* 17, No. 2, 265–278, DOI: 10.1080/14747731.2019.1652464.

Wittman, H. 2015. From protest to policy. The challenges of institutionalizing food sovereignty. *CFS/RCÉA* 2, No. 2, 174–182. DOI: 10.15353/cfs-rcea.v2i2.99.

Wittman, H.; Desmarais, A.A.; Wiebe, N. 2010. The Origins & Potential of Food Sovereignty. In Wittman, H., Desmarais, A. A., Wiebe, N. (Eds.): *Food Sovereignty. Reconnecting Food, Nature and Community.* Oakland, California: Fernwood Publishing, 1–32.

Zabalaza 2019. A ZACF anarchist in the Landless People's Movement, South Africa: Interview with Lekhetho Mtetwa. https://zabalaza.net/2019/04/05/a-zacf-anarchist-in-the-landless-peoples-movement-south-africa-interview-with-lekhetho-mtetwa/, Accessed on 24/02/21.

Zezza, A.; Tasciotti, L. 2010. Urban Agriculture, Poverty, and Food Security: Empirical Evidence from a Sample of Developing Countries. *Food Policy* 35, 265–273. DOI: 10.1016/j.foodpol.2010.04.007.

Zhan, S.; Scully, B. 2018. From South Africa to China. Land, migrant labor and the semi-proletarian thesis revisited. *The Journal of Peasant Studies* 45, No. 5-6, 1018–1038. DOI: 10.1080/03066150.2018.1474458.

3   Exposing marginalisation

Food and farming in the city

As part of the critical urban food perspective, this chapter uses the theoretical considerations on marginalisation in everyday life and related exclusionary dynamics as the point of departure. The task of this chapter is to present the KEF initiative, particularly its initial mobilisation, ambitions, the involved communities, related farming practices, interventions, and the organisational structure. Against this background, this chapter shows how members expose inequalities. Members' deprived livelihoods and restricted food choices fuel their ambitions and interventions, which refers to the notion of perceived space. In addition, historical characteristics of the involved neighbourhoods, Blanco, Pacaltsdorp, and Thembalethu, shape their activism and food production practices, mainly backyard gardening and to a lesser extent squatter farming. The production of space under the right to the city depends on collective power to remake citizens and the city (Harvey 2008, 23). As outlined before, various features in South Africa's commercialised agri-food system have negatively affected urban dwellers and those at the rural–urban interface. Alternatives proposed by the initiative are in the focus of the next chapter. In this way, their engagement with food sovereignty can be traced. The chapter uses insights provided in key informant interviews, focus groups discussions, neighbourhood walks, and participant observations. Newspaper articles provided further important information.

Initial mobilisation

The KEF initiative started in late 2012. The driving force of the initiative was one doctor and one of the cancer patients at the Blanco clinic. Both were deeply concerned about the nutritional deficiencies in poor communities. The doctor was mainly working in the field of chronic diseases – diabetes, TB, HIV, hypotension, and described the situation in Blanco, a rather poor neighbourhood, in the following way: "I identified one of the main issues as food security. [...] more specifically, it's a double whammy of malnutrition on the one side and obesity on the other side".[1] It was highlighted in the discussion that cancer can be driven by unhealthy nutrition and chronic diseases. In this context, the doctor thought further exchange with nurses and patients to create more sensitivity. She and one of the patients wanted to make a difference in in boosting the immune system through diverse and nutritious diets. Alex, the cancer patient, turned out to be a very committed

DOI: 10.4324/9781003182634-3

community member, experienced in gardening, and a member of the Wildlife and Environment Society South Africa (WESSA). Together, they started a garden on the hospital grounds for the patients, hospital staff, and visitors. In this way, they began promoting access to fresh and healthy produce and encouraged people to grow their own vegetables.

Beyond the hospital garden, they initiated a gardeners' group to reach out to the wider community. Many of the residents of Blanco have access to small patches of land and thus had the resources at their disposal to join the initiative. Brenda, a sort of community leader, networker, and passionate gardener in Blanco, joined them and attracted further people. She moreover built the link to gardeners who were previously engaged in another community gardening group called the Blanco Droom,[2] which was mostly focusing on beautifying gardens. These gardeners were also keen to step up and push food security. Further details on Blanco are provided in the following part of this chapter.

In one of the first meetings, they agreed on a name for their group. The local doctor remembered: "together with the group we brainstormed firstly on the name [...]. Kos en Fynbos stands for food security and biodiversity, they complement each other and that's what community resilience is built of".[3] Another member framed the motivation of the group in the following way, "in a place like George and maybe even in most of South Africa, we have the sun, the rain, the land, it should be easy for people to pull themselves out of a situation with no food or very bad food and quite easily change that around into something better".[4] The assumption was that gardening is one of the most viable ways of easily changing the situation of marginalisation. For many members, urban agriculture has become an integral element of social reproduction. Beyond that, the group activities are described as meaningful leisure time. Overall, the broader idea of self-help motivated many to join and to support the initiative.

In late 2013, two professors of the agricultural department of the George campus of the Nelson Mandela Metropolitan University (NMMU) joined and further shaped the initiative. They had previously been organic farmers in Natal, Zambia, and Mozambique. One of them suggested a community gardening competition to reach out and connect the gardeners in Blanco and maybe even beyond. The idea was to visit different gardens in a group with the participants and exchange experiences. He even donated money for prizes (e.g. wheelbarrows, hoes, and forks). As well as encouraging people in gardening, it was the intention to bring the people in the community together, and to step beyond individualised practices of buying, growing, or eating food.

At this early stage, the public competition event attracted many people and particularly benefited from Brenda, the community worker, who maintained strong connections to many individuals and turned out to be a brilliant organiser. One member described the process of inviting people as follows,

> She [Brenda] personally walked around all the streets and where she saw people with good veggies growing in their gardens [...] she dropped a competition form under the front door and asked the neighbours. [...] We ended up with 45 people who were competing for the prizes.[5]

The event took place in March 2014. The local doctor recalled: "The first year we had 50 gardens and the second year we had 400 gardens. So, it really just went".[6] Through the gardening competitions, people came together, saw that gardening is both doable and rewarding, and shared their motivations and knowledge.

In the preparation phase of the first competition, the municipality heard about the initiative and offered its support in hosting the prize giving and for future competitions. In addition, the *George Herald* offered to publish half a page every week about the initiative, thereby becoming the local media partner. In this way and through further competitions, the initiative gained visibility beyond the Blanco neighbourhood. Gardening competitions[7] were organised in cooperation with the municipality, WESSA, and the Landmark Foundation. Various sponsors were keen to fund prizes, which mostly comprised gardening equipment, seeds, trees, and plants. In this way, future collaborations were established. These donors and supporters are discussed in section "Organisation and cooperation" in this chapter. The related events starting with the first competition and further developments are covered in the *George Herald*.

Gardening demonstrations on no-dig gardening and composting were another important element of the initiative. Those attracted many people and helped to introduce KEF in different areas and to get people on board. Demonstrations were mostly facilitated by some of the experienced gardeners and farmers. The common starting point was the establishment of a no-dig or so-called lasagne gardens[8] which refers to the permaculture method. These beds are introduced on the top of the soil; leaves, weeds, or grass do not have to be removed beforehand, and the soil is not turned over (i.e. no digging). The first layer is usually cardboard or newspaper. The different layers set up on top are wet cardboard, wet newspaper, dry grass, fresh soil (which might include a bit of fresh compost, kitchen waste, and manure). These green, brown, and paper layers are put on top of each other. In this way, the bed contains organic material which will compost over time. Seedlings are planted on top. A detailed explanation of no-dig-gardening is provided by one of the members in the *George Herald*. The demonstrators and gardeners brought the required material (e.g. cardboards, dry grass). In this way, gardening is possible even if the ground soil quality is low and time in maintaining the garden is limited – no tilling and weeding are required. The main intention was to show that gardening is doable for everyone and to share experience and knowledge. Inspired by the idea of permaculture, one member emphasised "sharing is one of the major components of permaculture. The whole permaculture concept is to care for the earth and care for the people and share, share, share".[9]

While the demonstrator basically led the session, everyone was welcome to ask questions, share experiences and advice. Moreover, these workshops were "hands on" meaning everyone became active in building the no-dig bed and the compost heap. People who joined often brought something to eat or tea. Sessions were easily initiated by interested persons, a member elucidated, "So people just say, we would like to have a garden in our area, we would like to have a demonstration in our area. We make arrangements and then we go there, and then we see what we can do".[10] The communication on eye-level and the interest in involving everyone also made it easy to overcome language barriers. Although most of the KEF members speak Afrikaans[11] the dominant language in George, some in the group only speak English and people from Thembalethu mostly speak isiXhosa. Already in the first session, a mode of translating, switching languages or slowing down the explanations was established.

One of the founders described the spread of the initiative to other neighbourhoods and encouraging further people in the following way:

> At a certain point, we managed to bring a lot of people together. [...] This is what we did in Pacaltsdorp, too. Almost every week there was a demo. Usually. an older person invited all the friends and neighbours [...] Everybody was grassroots level, and everybody helped everybody else.[12]

Besides private gardens, these demonstrations took place on public and institutional grounds in different areas of the city (e.g. at the Cancer Society, George Hospital, Harry Comey TB Hostel, Ilingilethu Crèche). The demonstrations became a new kind of routine in the communities, carving out new ways of connecting inhabitants, growing, and thus accessing and eating food. From a critical urban food perspective, it can be clearly highlighted that these are simple elements in everyday life which can make a difference and overcome old routines and inherent social exclusion (cf. Gardiner 2004; Lefebvre 1991 [1947]).

Another reason for the spread of the group is its interaction at a personal level and its openness to everyone independent of their skills, resources, or garden size:

> We learn from each other, we listen to each other, we see what people have got. Some people have a tiny garden. Other people have a bigger place, it is all up to what there is really. [...] It is not like oh well can't have it because my garden or my circumstances are not right.[13]

Most of the members are growing vegetables and herbs for themselves in their backyards. Spaces vary between 4 and 10 square metres. In addition, small door gardens can be found, including so-called door gardens which are only the size of a door. KEF experience shows that a high variety of vegetables can be planted on this small plot of land with only little workload. A few members are cultivating on fallow land at the fringes of Thembalethu and sell part of their produce informally. Others are maintaining gardens on public grounds (e.g. in schools, crèches, hospitals). This is further elucidated in the following part.

Other communities gradually joined KEF. As indicated in the quote above, the suburb Pacaltsdorp (comprising Rosedale and New Dawn Park) became the second community involved in KEF. The municipality encouraged the initiative to reach out to Touwsranten, which is far from the centre. The Local Economic Development Unit (LED) and the NMMU had previously tried to push urban agriculture there but failed (LED 2014). Further efforts via competitions run by the LED brought the communities of Rosemoor, Le Vallia, Protea Park, Rosedale, Kleinkrantz, and New Dawn Park on board in late 2015 (LED 2016). These communities are included in the map in Chapter 2. In early 2016, Thembalethu joined, although it was not involved in the competitions earlier. Members emphasised that KEF became an important connecting point for these previously separated communities and exposes them to other realities beyond their neighbourhood. On gardener remembered

> because of the divided past a lot of South Africans don't actually know how other South Africans live. I think what KEF actually does is opening each other's eyes up. [...] It was very interesting for us to see when we were in Thembalethu the coloured KEF members from Blanco and Pacaltsdorp.[14]

Looking back, particularly members from Blanco, Pacaltsdorp, and Thembalethu were driving forces and strongly shaped the initiative. Representatives of these areas were always to be found in the regular meetings and KEF events during the main research period and thus receive specific attention in this work. The activities in the other communities are rather driven by the interventions of the municipality. A take-away message from this introductory section is that the mobilisation of the initiative is driven by different factors: starting with health issues rooted in restricted access to quality food and at the same time diverse opportunities to grow at least a bit of fresh food for instance on private or public land using available (shared) resources such as knowledge, tools, and seeds in local communities. To better understand these local mobilisations, it is key to have a closer at the roots of the participants in the broader socio-economic dynamics of their neighbourhoods.

Socio-economic backgrounds

At the beginning, a small group of well-educated individuals (e.g. doctor, professors, NGO workers) were important initiators of the initiative. Besides a strong desire to lift-up the poor, their professional life built a bridge to their engagement. However, some of them moved away in the first years. Brenda, an elderly coloured lady and well-known by everyone throughout her profession as a passionate community worker in Blanco, opened the initiative up to the wider community of mostly deprived people. The majority of the KEF members are coloured, with a smaller number of black (mainly from Thembalethu) and white people.

In early 2016, one of the founders indicated that more than 700 individuals were active in KEF. Table 3.1 provides the estimated numbers of members in the focus communities, Blanco, Pacaltsdorp, and Thembalethu, as well as some additional details about the respective areas.

The settlements involved in the initiative are strongly characterised by poverty, which is also shown by the high number of dwellers without a monthly income (Table 3.1). This precarious income situation correlates with a high risk of hunger. For the census of 2011, "the lower bound poverty line for March 2011 was set at R443 (per capita, inflation adjusted poverty line) meaning that any individual earning less than R443 a month would have to sacrifice essential food items in order to obtain non-food goods" (Western Cape Government Provincial Treasury 2015, 13). This implies that people do not only suffer from income poverty, but rather from multidimensional poverty (food poverty, education poverty, etc.). One of the interviewees summarised: "It's mostly poor people. Not all of them but

Table 3.1 Overview of the three focus communities

Suburb (Wards)	Dominant Language	Population and estimated KEF members	Composition of population[15]	Income Poverty
Blanco (Ward 1)	Afrikaans	9,350 320	66.6% Coloured 24.6% White 7.1% Black African 1% Other 0.4% Asian/ Indian	37% of the population are without monthly income 33.5% of individuals earn R1–R3.200 per month
Pacaltsdorp (Ward 14)	Afrikaans	9,571 200	94.5% Coloured 4.5% Black African 0.6% White 0.3% Asian/ Indian 0.1% Other	41.5% of the population are without monthly income 29.7% of individuals earn R1–R3.200 per month
Thembalethu (Wards 9, 10, 11, 12, 13, 15, 21)	IsiXhosa	43,103 (estimation 90,000[16]) 60	91.8% Black African 5.5% Coloured 1.6% White 0.7% Other 0.2% Asian/ Indian	40.5% of the population are without monthly income 42.16% of individuals earn R1–R3.200 per month

Sources: George Municipality 2014, Statistics South Africa 2011 and own investigation 2016.

many poor people, who can't afford to buy vegetables are involved".[17] In this way, food cultivation has become an integral part of social reproduction, tying in with Shivji's notion of the working people and dynamics of semi-proletarianisation (2017). Using the critical urban food perspective, it can be argued that those directly in need engage in urban agriculture.

In South Africa, about 45% of households rely on social grants, mainly (Foster) Child Support Grant, Older Person's Grant, Disability Grant, and Care Dependency Grant (Statistics South Africa 2018). While there are no precise numbers available for social welfare in the target communities, the exchange with KEF members revealed that social grants are an integral income category. A member from Pacaltsdorp mentioned: "child and elderly grants and pensions keep many KEF households going. Everything else is uncertain".[18] The latter part of this quote refers to unemployment and precarious employment situations. A social worker in Thembalethu estimated that more than half of the KEF members are "unemployed or make some kind of informal income".[19] Many combine different livelihood strategies, from jobs in the informal sector to food production for household consumption. Moreover, individuals with little or no monthly income rely on wider safety nets in their households and communities. A small number of people – mainly part-time farmers from Thembalethu – sell their produce informally to local shops, small restaurants or food stalls. A member from Thembalethu summarised, "Not many of us sell, but I regularly bring my stuff to town, mainly spinach. It's not more than R50 a week").[20] She has an informal part-time job as a cleaning lady. Further details are provided in the brief microhistories in the following.

The two KEF leaders who were appointed in 2016 illustrate the diversity of the initiative's members. Catherine had been working for the Landmark Foundation, an environmental NGO. She describes herself as an environmental activist; she has been a passionate gardener and community educator. She is one of the few vmembers combining her jobs with KEF activities. For instance, those working in environmental NGOs, social work, or health care use their position for community outreach activities. Sipho was the previous KEF coordinator in Thembalethu; he is the representative of the Farmers Union Umanyano and part-time farmer in Thembalethu. Although they represent different segments of the working class, their lived experiences show that they converge in their struggle for a better food system.

Different types of urban agriculture and microhistories

The following sections provide details on the three main suburbs involved in the KEF initiative, related activities, and food production in these areas. Although Blanco, Pacaltsdorp, and Thembalethu are part of the Greater George area, they are unique in character. They were previously independent municipalities, and all three are still considered separately in spatial planning (with separate plans). In 1995, under the Local Government Transition Act, the City of George amalgamated the local governments of Thembalethu and Pacaltsdorp, which were formerly separate black and coloured local authorities. Today, the village character of Pacaltsdorp, Blanco, and Thembalethu is still supported by large areas of empty fields and streams at the borders of these neighbourhoods. Some of the spatial barriers which demarcate these suburbs are so-called buffer zones.

The government introduced buffer zones to separate different population groups with the purpose of minimising racial interaction during the apartheid era. Whites had the privilege of staying in large central areas, while coloured and blacks were displaced to distant urban periphery townships. Although apartheid ended in 1991, in many South African towns these zones are still fallow land. According to the South African Cities Network (SACN), "the size of these urban places is larger than might normally be expected for centres so closely spaced" (2014, 78). In some of these peripheral areas, farming activities can be found. However, in several discussions, it became obvious that there is a strong desire on the part of the government to utilise these open spaces and former farmlands surrounding the city for residential development (e.g. gated communities). There are also ideas to use it for eco-tourism. Food gardeners in the areas have been keen to work this land. For instance, squatter farmers can be found at the fringes of Thembalethu. However, KEF's intentions to utilise the buffer zone in Blanco and the Old Crocodile Farm at the outskirts of Pacaltsdorp were rejected.

Besides limited and small interventions in the buffer zones, KEF food producers mostly cultivate backyard and door gardens in all three areas. The initiative truly presents a variety in food production practices and crops. In the old neighbourhoods of Blanco and Pacaltsdorp, some gardeners even have fruit trees. The most common trees are avocado, citrus, fig, and apricot trees. In terms of vegetables, people mostly grow onions, garlic, cabbage, beans, tomatoes, spinach, carrots, sweet peppers, sweet potatoes, lettuce, chillies, beetroot, cucumber, pumpkin, and peppadews. Sweet potatoes and pumpkin, particularly, are considered traditional crops. In many gardens, herbs can be found. Well-established gardens also feature banana plants, berries, maize, broccoli, cauliflower, Brussel sprout, and asparagus. In general, gardens are mostly irrigated using tap water, standpipes, or by carrying water in buckets filled from stand pipes or streams. Some gardeners also harvest rainwater.

In Pacaltsdorp and Blanco, the majority of people grow food in backyards ranging from approximately 4 to 10 square metres. Input and output vary widely and are difficult to quantify. Many gardens are maintained by individuals. One gardener from Pacaltsdorp who works the soil with her daughter and husband proudly stated that they harvested "at least 10 kg sweet potatoes, maybe three buckets of green peppers, six buckets of apples and several maize plants" in the last season.[21] They also had smaller amounts of tomatoes, onions, berries, and salad. But she also admitted that many gardeners grow similar things, mostly harvest at the same time, and storage can be challenging. In both areas about a quarter of members started gardening in smaller door gardens, mainly salad, onions, spinach, and herbs. In Thembalethu, a densely built settlement, people have smaller backyards, a maximum of 5 square metres. The majority grow food in tiny door gardens. In addition, a small group of squatter farmers grows vegetables at the fringes of Thembalethu.

Apart from the individual gardens, there are also some community gardens; according to one of the founders: "There is one at the Botanical Garden, there is one at the cancer rehab house, there is one in Pacaltsdorp highschool [...]. Blanco wants to establish a community garden. But it is having difficulties with municipality regarding land in the buffer zone. There are community gardens in Thembalethu but it is more like a coop".[22]

Reinforced by diverse farming and gardening activities, an old village charm prevails in Pacaltsdorp and Blanco, which are also home to a number of historical buildings. Thembalethu gives the impression of a typical township on the move characterised by the steady influx of new arrivals. Nevertheless, they all share a story of historical exclusion and marginalisation. According to the SACN these are "less well-off areas on the periphery of the old town, which mainly serve as dormant neighbourhoods with little economic opportunities, namely: the older settlements of Blanco and Pacaltsdorp [..., and] the newer area of Thembalethu" (2014, 84). It is evident from the stories of the research participants that historically people from these communities experienced different kind of oppression and trauma. One of the interviewees described the differences between these areas in the following way,

> In Thembalethu people are poorer than in coloured areas and also they have been messed around a bit more [...]. In KEF we've got people in the old areas, like in Blanco, where they have been basically growing food for generations and they have fruit trees [...] food is on a different scale.[23]

Interviewees highlighted that black people in the younger settlement of Them-balethu are apparently often struggling to put enough food on the table as there is not much extra.

The character of the three areas is outlined in the following sections. In fact, these areas demonstrate the interplay of urban and rural agrarian characteristics, the so-called rural–urban intersection, which is emphasised in the critical urban food perspective.

Blanco

The suburb of Blanco is located in the north-eastern part of the city. Until 1974 Blanco was a separate local authority and farming village. Due to its growth, it became part of the George Municipality. Under apartheid, it was declared a "coloureds only" area. Today, Blanco is more mixed in population composition in comparison to other suburbs in George (see Table 2.1). Some of the interviewees praised the area for its "old village charm", as "a nice nuclear community that has a sense of community" and "the community's language is very homogenous".[24] Another research participant added, "Blanco is a community where people have been planting their own food for generations. It is a very old community".[25] This is one of the reasons why KEF became popular in Blanco. Most of the houses and small dwellings sit on small plots with a bit of space for gardens.

However, quite a number of backyards are occupied by informal dwellings. According to the 2011 census, more than 9.1% of the households are based in shacks[26] (Statistics South Africa 2011). Before the overthrow of the apartheid regime, several informal structures were evident at the fringes of the ward (e.g. the Oubus settlement, whose residents were evicted to Urbansville). In addition, several middle-class houses without fences around them can be found, which underlines the rural character and the related perceived safety.

In the past, Blanco was well-known for its hop plantations, and George is still the only area in the country where this essential beer ingredient is grown. Some of the residents of Blanco worked with a local food-producing company in the area, which closed a couple of years ago. Most of the community members have roots in agriculture – farming and gardening have been integral to the area for generations. One interviewee from Blanco pointed to the renewed interest in urban agriculture in the area:

> I'm from Blanco; it's a village like suburb. You know, where I grew up we were in a very big garden. We had more than 300 chickens. I can't recall going to the shop to buy meat, vegs or fruits, because we had everything in the garden. And now after so many years I am back there. That's a nice thing.[27]

Today, most of the agricultural land is privately owned. Many people would be interested in using smaller pieces for subsistence farming: discussions with the municipality are ongoing. Besides vegetable and fruit cultivation, some of the KEF gardeners also raise chickens. One member even has a donkey cart for transport. The flower lovers amongst the gardeners are particularly proud of their indigenous flow-ers (such as lilies). Before KEF, many of them were members of the gardeners group

Blanco Droom. The coordinators estimate that KEF has 320 members in Blanco. In November 2015, more than 300 gardens entered the gardening competition.

At fringes of Blanco, Lucas has a typical verdant garden covering a bit more than 6 square metres (cf. Siebert 2020). Behind the house of his family and next to the shallow stream, Lucas set up diverse and ecologically rich flower and vegetable patches. Diverse crops including herbs, tomatoes, cucumber, and eggplants can be found. His garden is inspired by the ideas of permaculture and companion planting, which is particularly visible in the composition of plants. During the neighbourhood walk, he proudly presented stalks of maize. These spend shade and improve growth of potatoes and beans. Neighbours appreciate that he is always willing to share seeds. Following the overall KEF philosophy, he rejects hybrid seeds and prefers homemade organic fertiliser. For many people in Blanco, he is the ideal contact person for KEF; he is keen to offer support and guidance whenever needed. In the discussions, he highlighted often, that a garden implies independence, appreciation of traditional knowledge and thus contributes to health. For instance, he mentioned once: "the old people said their health is coming out of their garden. Use your own veggies and your own herbs".[28] His household strongly relies on his backyard food production. Still, he is in the position to share surplus frequently including a certain portion he sells at a local house shop. In addition to that, the household combines pensions and social grants (mainly for older persons) and the meagre incomes of relatives. Lucas himself has become active in paid gardening work in affluent neighbourhoods. Such diversified livelihood strategies are prevalent amongst the members of the initiative.

Pacaltsdorp

This suburb is located in the south of the city centre. Like Blanco, Pacaltsdorp was previously a village. One interviewee described Pacaltsdorp in the following way,

> It was a mission station – it's really got a very, very long history. People there consider themselves Pacaltsdorper. They are not from George. And what you find there is a bit of a rural feeling. Larger portions of land often historically held by families. [...] I think lower density like you find in places like Thembalethu.[29]

About 7% of the households live in informal dwellings mostly in backyards (Statistics South Africa 2011). Pacaltsdorp also includes large new governmental housing developments like Rosedale; however, this is an infill development of more than 900 houses which is spatially rather separated from the rest of the suburb. A representative of the municipality admits, "That is really a forced sort of community. You know, you got a house, you moved in, there you are".[30]

Like in Blanco, many of the residents would like to use some of the privately owned land at the fringes of the settlement for subsistence-farming. Again, land access is difficult. The municipality is aware of the situation and considers making smaller pieces of land available via a tender process. Challenged by high unemployment rates, the LED, George Municipality, and provincial Department of Agriculture, were considering establishing an Integrated District AgriPark including a vegetable drying facility to create new jobs and economic growth. This was also related to the shutdown of the McCain vegetable processing plant in the southern part of the city (with about 140 employees, some of them only seasonal employees, and about 40 farmers who supplied the company). A few of the KEF members indicated that they or family members had worked there. In 2016, the LED mentioned that the AgriPark plans were difficult to realise and an AgriPark has already been built at Oudtshoorn, a neighbouring town.[31] Still, the deputy director of the Farmer Support and Development, Department of Agriculture, indicated that the AgriPark is still on their agenda.[32] In several earlier discussions, it became evident, that the new AgriPark plans sparked hope amongst many Pacaltsdorpers for employment but also for the opportunity to sell their harvest there.

The gardens are quite similar to those of the Blanco members. In 2016, many Pacaltsdorpers started with small door gardens. Lilian was one of them. I visited her several times. She first invited the group to her house for a gardening demonstration. They set up a lasagne garden and a compost hub on the lawn in front of her house. She admitted that gardening has been a part of her family; this is mirrored by many other Pacaltsdorpers. However, she indicated that she and her husband have been employed in the local retail sector over the last 40 years, so it is only now that she felt she had time for gardening. Because of health reasons, they both work less, and thus "free food provision from the garden is indispensable".[33] Despite a number of novice gardeners, many others, more experienced food growers, have been using every centimetre of their backyard over the last decades.

Thembalethu

This township was proclaimed a separate municipality in 1986 and amalgamated into the George Municipality in 1995. Thembalethu, meaning "our hope" in isiXhosa language, is the largest township of George. Like many other townships in the country, this settlement has its origins in the apartheid era. The settlement has grown dramatically, particularly following the change of government after the end of apartheid. Since then, an ongoing transition of the area in terms of new arrivals, housing, and commercial developments with a strong informal sector can be observed. According to one interviewee, people in Thembalethu "don't have a sense of community because they have been shifted around and there is a lot of migration – in-country migration".[34]

Today, un(der)employment – including informal and non-permanent work – is still the sad reality. In addition, a lack of proper housing and basic service supply are still serious problems. After apartheid, cheap housing was constructed closer to the city centre as part of the government's Reconstruction and Development Programme. In addition, several simple (concrete), older houses can be found. The large majority of dwellings, however, are an ever-growing number of densely built informal shacks with barely serviced plots at the fringes of the quarter. An interviewee mentioned, "In Thembalethu, the people don't have a lot of space. So, should they want to plant a bit of veggies. Their houses and shacks are really on top of each other".[35] Land access for gardening and farming is definitely a problem in this area. This is described in more detail in the following chapter. Nevertheless, many aspects including large swathes of open fields at the fringes remind the visitor of a rural village. The description provided by Lanegran and Lanegran still sums up the areas' character perfectly: "Although they are high density settlements, they are not completely urban. While some have municipal services, most do not. [...] The areas are homes to litters of pigs, flocks of chickens, small herds of cows, sheep, goats and the occasional donkey and horse. These animals roam freely through the communities" (2001, 679–680). These experiences show the restricted availability of land for livestock farming.

As highlighted in some of the protocols, the suburb Thembalethu became involved in KEF driven by passionate and well-linked area coordinators and through several gardening demonstrations in early 2016. Mainly backyard gardeners became involved. Zandiswa is one of the KEF members in the area. She lives at the outskirts of the bursting township. Although her farm plot has a similar size like Lucas's garden, her family's life is completely different (cf. Siebert 2020). Together with a few friends, they have been occupying land and living in shacks since 2014. Still, she is concerned about eviction but maintains "Municipality cannot send us away".[36] Driven by optimism and KEF's work, she continues to grow vegetables on no-dig patches and even has a few chickens. Her neighbours' cows are roaming around freely in the area; here she gets a bit of mild in exchange for eggs. She sells part of her produce to a coffee shop in town. In addition, the household relies on her informal job as a cleaning lady and on child grants.

Thembalthu's rural character is supported by the wide farmland called Sandkraal between the settlement in the north and the coast of the Indian Ocean in the south. In mid-2016, 14 squatter farmers active on the land of the former Sandkraal engaged with KEF and thus added a new notion to the initiative. Historically, many of them were connected to the land and agriculture in their villages. One KEF member highlighted that many of the small-scale food producers "were farm workers. Not only people in the old areas, like in Blanco have been growing food for generations".[37] Most had migrated from the Eastern Cape. However, squatter farming activities at the outskirts of Thembalethu are very precarious, characterised by the absence of official land access, restricted storage and water access. They use about 10 square metres each and mostly grow cabbage, spinach, onions, and occasionally potatoes and pumpkins. One of them explained "We rely on our hands" and told me proudly, "I got 30 kg of cabbage [in the last season]", which he sold informally at the Thembalethu taxi rank.[38] Despite a bit of income through farming, they rely on additional income activities. Some of these squatters are part of the Themablethu's Farmers' Union Umanyano, which seeks to mobilise and represent farmers' interests in official land access in the area. It is represented in KEF through its leader Sipho.

Organisation and cooperation

Between 2012 and early 2016, the KEF group had no official leadership body. Those who were considered as leaders – directly related to their role as initiators – rejected this role. However, the growth of the initiative increased the requests for official leaders not only from inside but especially from outside (governmental side). It was only in 2016, that KEF elected leaders in a meeting. The election was carried out by secret ballot, and Sipho and Catherine were elected.

As well as electing leaders, the early cooperation with the municipality and the spread of the initiative called for area coordinators: "And they [the municipality] decided that a good thing to do would be to have area coordinators for the different areas as it grew. So, Blanco has two area coordinators and Pacaltsdorp had a few, Touwsranten and blabla that will be remunerated by the municipality for their coordinating work".[39] The communities and those who were very active in the initiative decided who would be the coordinators. The coordinators' role has mainly been to share information regarding KEF's work, to provide mutual support, and to be the first point of contact for the community members and the municipality. While the municipality and the Department of Agriculture often worked through part-time external service providers, local community coordinators showed that they knew their neighbourhood including the people, the circumstances and the location quite well (LED Project Plan 2016, 5). Over the years, it became clear that the character of these coordinators also pushed the group further, as one member summarised:

> It is very much individual-based too. If we have a person in an area like auntie [Brenda] in Blanco. She inspires people around her. So, it becomes a different thing there than it will become in a place where […] the contact person isn't very involved.[40]

Throughout the period of cooperation with the municipality, mainly the LED, the coordinators were remunerated and were asked to visit other members and approach potential new gardeners twice a week. It was mostly persons with experience in gardening who were asked to become coordinators; some of them had even won a gardening competition previously. The coordinators kept track of those interested in competitions and gardening demonstrations. They also joined the regular meetings to exchange information, share the news of their community and get insights from others for their community. A coordinator from Thembalethu described her responsibilities in the following way:

> you invite a person to the group. As soon as this person is in the group, it will get the experience, immediately that garden will now prosper. So, the neighbours will see. […] when he starts get interested, invite him or her to the group as well. So, we are planning to see this gardening thing growing.[41]

Despite the appointment of leaders and coordinators, it is important to stress that the group is not hierarchical in its structure and the role of the leaders is rather for representative purposes. Although there are members who take the lead in certain activities, e.g. coordinating meetings and demonstrating gardening practices, the initiative operates on the basis of an equal footing among members. Some even referred to it as family and close friends. This is another integral factor which makes the group attractive and easily accessible for many. It is considered a "community outreach initiative", meaning that members take care of one another and motivate each other. In this way, the group does an important job integrating those who suffered from exclusion at different stages, for instance through racial segregation, or in terms of education or long-term employment. In contrast, KEF creates pride, appreciation for knowledge, ties to farming, and hence food production skills.

The regular group meetings have brought the coordinators and interested members together. Decisions are mostly made in these meetings. One of the KEF founders described the meetings in this way: "There is a meeting every second Wednesday of the month. They are usually either held at the municipal buildings because they allow us access to their buildings or at the George Museum or at the Botanical Garden".[42] The gatherings became a vital platform to strategise and exchange seeds, produce, and plants. It was also decided in the meetings to have certificates for the members.

WhatsApp is the main communication medium for members across different communities. The coordinators and all those who are interested are connected in a WhatsApp group. Meetings and events, interesting information (recipes, advice, jokes, and motivational words) or simply photos of a successful harvest are shared and celebrated there. However, those who do not have a cell phone or WhatsApp sometimes struggle to get in contact or hear the latest news. Coordinators try to keep these people updated. Newcomers sometimes contact the municipality if they do not know how to get in touch with KEF. The *George Herald* intends to prevent from these difficulties. Therefore, its newspaper articles on KEF regularly include contact details of the area coordinators for interested people and organisations. It is not only new members who get in touch, but also sponsors, donors or other groups who would like to cooperate.

Besides its rapid increase in members, the work and outreach activities of the initiative have attracted many cooperation partners and supporters in the area. Some of them have had a role in shaping KEF's work and mission, others support the initiative, for instance through donations in the form of seeds or prizes for the competitions. The wide support which the group attracted encouraged the members to continue and step up their work. It has to be stressed that none of these organisations explicitly engages with food sovereignty. However, food access and biodiversity related to community support, knowledge creation, and implementation of urban agriculture are part of the work of some organisations. These actors are briefly introduced in this section, in alphabetical order.

Garden Route Botanical Garden

In George, the Botanical Garden is a tourist attraction and a recreational space. Besides the activities of the garden itself, it is also a registered NGO which is active in environmental protection and community work. One of the KEF founders describes the initial cooperation with the botanical garden in the following way,

> we asked them if we are able to use their administrative capacity because we found it too early to register an NGO [...] and we didn't want to duplicate because the botanical garden has as its aims also outreach to the community. We thought this could be a good vehicle for them to do outreach.[43]

As well as this administrative support and cooperation in community support, the botanical garden became the central meeting place of the initiative in 2015; it has also provided sponsorship for competitions and donated gardening materials. Some of the members have been working in the botanical garden, particularly the permaculture site. The garden hosts a number of events in summer such as concerts and fairs, and several of these have been organised together with KEF. For instance, in August 2014 the botanical garden hosted the National Science Week, and KEF was very active in supporting these activities. The initiative organised a traditional Khoisan breakfast and held permaculture demonstrations. Moreover, the botanical garden has a herbarium, which shared its knowledge with the members and hence supports the initiative.

George Herald

The local newspaper has been supporting the initiative since early 2013 and offered to publish one article per week about KEF to create more exposure and reach a broad public. One research participant explained: "George Herald started putting people in the newspaper who otherwise would never have featured in the newspaper. That is helpful, I think. That might make sense. It makes them proud of what they are".[44] Members usually write the draft text and provide photos, and the articles are published with only minor editing. From time to time, journalists join KEF activities themselves. The newspaper has proved to be very supportive, which provides a unique opportunity for KEF to reach out to many readers. Sometimes KEF shares these articles on its Facebook page. Regularly, coordinators of different areas have shared their contact details and meetings have been announced in the newspaper. The *George Herald* is published weekly as a print edition. In addition, articles are published daily in the online version. News on KEF is published in both versions. Journalists have also been active in sharing relevant contacts with KEF and building linkages to other initiatives working in the broader fields of social and environmental sustainability. For instance, the Avo's for Life project, initiated by local churches, contacted the *George Herald* in an attempt to reach out to the wider communities in 2016. The newspaper linked the project with KEF, which became a driving force in planting avocado trees in the area. In addition, the *George Herald* has provided support to competitions in terms of donations, e.g. seeds.

George Municipality – Local Economic Development Unit

The George Municipality, particularly the Local Economic Development Unit, has been interested in the community-led initiative right from the beginning. First, the LED gave its support through donations, e.g. water tanks, seeds, and trees, as part of the first competition. The municipality also hosted the prize-giving ceremony. Afterwards, the LED became involved in initiating and funding the competitions, which it sees as related to its own mission. An expert of the LED defined the role of the unit in the following way, "creating an enabling environment for the local economy to grow. [...] LED is about local collaborations, local stakeholders all collaborating to use the local resources, to grow and make a more sustainable place for sustainable futures, and there is no specific way to do it".[45] While LED's overall focus is on economic growth in close relation to the Integrated Development Plan for the area, the interviewee also pointed to strengthening the people as key actors in the economy,

> If the local people don't have the skills or the will it is challenging. Initially, we were very competitiveness focused but over the last five years it has evolved to be a more balanced approach to do both sector development strategic work and to do lower level community work.[46]

One of the local economic development plans is to create more employment opportunities including the establishment of agro-based enterprises (SACN 2014, 65–66). In 2015 and early 2016, the LED was interested in implementing an agri-park in the area, where locally produced vegetables and fruits could be processed or dried. It was eventually decided at the district level that the plant should be built in neighbouring Oudtshoorn.

Over the last decade, the LED targeted marginalised communities through several household food security projects, which mostly featured the implementation of food gardens. Under the Economic Development Strategy these were considered to be "quick win projects" (LED 2016, 3–4). In this vein, successful food gardens were seen as entry points for poor people into the economy. The overarching objective of these projects was "to create a culture of sustainable urban food production in George" (ibid. 6). The four sub-objectives pointed to: increasing the number of households involved in small-scale urban food production to improve nutrition levels; building relationships between the communities and the municipality; creating entrepreneurs through offering opportunities to sell surplus produce; and contributing to skills development, i.e. horticultural skills (ibid. 7). There is a clear link here to the work of KEF, which in its broadest sense intends to lift up communities and to improve gardening skills. While the LED previously failed to achieve the long-term commitment of the communities, it was inspired by the KEF initiative, which managed to get many people on board on a voluntary basis (LED Report 2014). The top-down approach of the LED and short project durations (three to four months with follow-up phases at a later point) restricted the impact of its activities. Both parties identified many parallels in their aim of improving food security. However, over the course of several jointly organised gardening competitions and the introduction of KEF coordinators, many members agreed that the municipality undermines the idealism of KEF. While the municipality supported the initiative in some ways, it also restricted KEF by blocking land access for community gardens on public land and farmland in townships and distributing monetary incentives to successful food producers. These developments are outlined and analysed in Chapter 4, Section "Access to land".

The LED's food gardening efforts have been closely aligned with the provincial Department of Agriculture, which provides support for gardeners and smallholders upon application, and which previously sponsored prizes for several competitions.

Landmark Foundation

The Landmark Foundation is a South African conservation NGO, based in George, which focused on biodiversity and sustainable development. Catherine, one of the KEF leaders elected in 2016, works with the Landmark Foundation in the field of environmental education, particularly in schools. Her background and skills have been essential in shaping KEF's community outreach activities in terms of gardening demonstrations. As part of the Landmark Foundation's work, she managed to introduce the KEF idea to other communities in and outside of George. One of the members described this spread in the following way, "She then took the concept to the countryside [small communities outside George]. […] all of that area has got KEF gardens in preschools".[47] As part of the Landmark Foundation's educational and community outreach activities, Catherine transports manure, grass, cardboards, seedlings, and other materials required in the gardens to different communities and schools. The Foundation has also supported gardening competitions with prizes.

Nelson Mandela Metropolitan University, George Campus, Saasveld

Two professors of the School of Natural Resource Management were important driving forces of the initiative, helping to initiate the first competition and sponsoring prizes. They were keen to share their knowledge on permaculture and organic farming with the members. This included demonstrations and visits to the permaculture site at the university. One of the KEF leaders summarised the university's contribution in the following way:

> They were quite happy to provide skills and knowledge transfer. So, they ran a workshop last time – a knowledge sharing workshop – very well organised. They did a compost workshop. This is how they support. I link up with them when if we think that we need their support.[48]

In August 2016, another professor and her students organised a KEF Social Learning workshop at the School of Natural Resource Management. The purpose of the workshop, which had 21 participants, was to provide an opportunity for KEF members to develop their social capital and share their tacit knowledge. This knowledge was used to create the "Kos en Fynbos Encyclopaedia of Gardening Knowledge". The NMMU thus plays an important role in sharing academic knowledge and it also values KEF's gardening practice and ideals. This kind of appreciation is perceived as a motivational force. NMMU representatives and KEF members told me that they are flexible in providing any kind of support needed by the initiative. These links with the local university have attracted local and national, as well as international, students and researchers in diverse fields (e.g. medicine, development studies, engineering) to engage with and learn from the local food producers.

Wildlife and Environment Society of South Africa (WESSA) in the Eden region

The NGO WESSA is a membership and donation-supported organisation with a focus on environmental and conservation projects in South Africa (Wildlife and Environment Society of South Africa 2019). Its tagline is "people caring for the earth". Some of the WESSA committee members have been very active in KEF and thus shaped its spirit regarding valuing and protecting nature. KEF leader Catherine, who has been a member of the committee, explained that:

This committee actually is just a committee under the umbrella of WESSA. [...] They do their own thing, bring awareness. [...] WESSA has committed themselves to 2000 Rands a year on seeds and whenever we need anything. They are always there. They are there to help us with everything. It is another partner.[49]

WESSA Eden collaborates with the Garden Route Botanical Garden in George and has its office there. The society has been active in the region for about 15 years, supporting a wide range of organisations and schools in the field of environmental education. WESSA donated seeds to KEF and co-hosted several public events with KEF such as the National Science Week in 2014. Many KEF events and achievements have been shared in the WESSA Eden newsletter.

Further actors: "moral supporters" and other donors

KEF introduced community gardens in many several institutions: hospitals (e.g. Blanco Clinic, TB Hospital Harry Comay), schools, and crèches in the involved communities, with the intention of using them as inspiration for the broader communities. However, it is not only members who regularly visit these sites; others also help to maintain the gardens and use these vegetable beds as community gardens. In this way, some of the clients have become involved in KEF. Some farmers who are within easy distance of George provide manure, and KEF members transport it to the different areas in town. Other organisations which have sponsored prizes for competitions are the Western Cape Government's Outeniqua Research Farm, Farmer Support and Development Unit (Department of Agriculture, Western Cape), the Links (Fancourt Golf Course), Genesis Seedlings, and the Plant Nursery. Permaculture South Africa, a local NGO, also provided sponsoring for the competitions and led demonstrations in the initial phase of the group. In addition, the George Museum has served several times as a meeting venue.

Overall, the initiative connects different groups of people and builds interlinkages across neighbourhoods, institutions, and organisations (e.g. NGOs, university, local newspaper). All this demonstrates KEF's intention to create a bigger network for a better food system. Different actors' wide support and cooperation imply the potential to strengthen the interests of the initiative. This ties in with O'Brien's idea of using "influential advocates" in resistance (1996, 36). For instance, the George Herald, which regularly publishes about KEF's work, follows up on the initiative's land requests and related work of government departments.

Critical reflection of the prevailing nutrition landscape

In the final part of this chapter, it is the task to carve out the perceived space and thus the conditions of the food system, particularly the shortcomings, which build the starting point of KEF's work. The motivation of the initiative to engage in food production can certainly be traced back to diverse reasons, which in part are highlighted in earlier sections including poverty, unstable integration in the labour market, and restricted, unhealthy nutrition landscape. Applying the critical urban food perspective, this book reveals that these reasons are part of a larger critical stance towards the prevailing food system in George, which is explored as a foundation to understand the initiative's interventions and the alternatives they are creating.

The initial steps of the group were made with the intention of providing access to fresh and healthy food and thus improving nutrition in poor communities. Beyond that, it became part of KEF's philosophy to work hand-in-hand with nature following the ideals of permaculture. To unify their mission and strengthen their vision, KEF members established a so-called creed in 2016, which they proudly show on their membership certificates and read out in a meeting:

> Kos en Fynbos is a ground-swell movement by people from our community who want to apply their passion for gardening to feed and beautify their homes, their community and their environment in a healthy manner. We support each other in reaching out to promote a functional, motivated society that cares for one another and shares in the abundance offered by food gardens. We care for the soil and for our fynbos heritage by promoting methods that can utilize these resources in our gardens in a sustainable manner. We don't expect anything in return as our prize is the inspiration that we give to others.[50]

This mirrors the movement's ecological and social values. Members told me that membership certificates are not about applying any pressure, but rather about strengthening the KEF message and sense of belonging. There are no strict conditions related to the certificate. Several members indicated that the creed is rather a simplified version of KEF's mission and vision, which began to further crystallise throughout several discussions and meetings.

The creed also points out, indirectly, that the movement abstains from monetary donations and opts for material support and sharing of skills as well as knowledge. This was also the motivation for KEF's decision to introduce a summer garden festival, which focuses on visiting well-maintained and inspiring gardens, instead of prize-oriented competitions. Today, gardening competitions remain a feature only of the municipality's projects. In this context, KEF had some struggles in 2015 and 2016 which were related to its collaboration with the George Municipality and the remuneration of some members. These tensions are explained in the following chapter. Several more statements and discussions show that the motivation and intentions evident in the lived realities of KEF members are much stronger than the creed. For instance, the *George Herald* stated, "the aim of KnF is to promote food security and a healthy environment through the growing, by permaculture methods, of organic and non-genetically modified, pesticide-laden food" (2015). Beyond this well-articulated aims in the newspaper and in the creed, lived realities show in more detail that the initiative reflects critically on the industrial food system, including the food offered in supermarkets and the dominance of food retailers.

Those who are involved with KEF have experienced the effects of a shrinking range and affordability of food, the domination of supermarkets, the transformation of dietary patterns, and health concerns. This links to a broader "crisis of social reproduction and the social wage" (Zhan and Scully 2018, 1023), and leads the movement to create and maintain local, informal structures for the production, exchange, and purchase of food. Several KEF members have experienced malnutrition and difficulties in proper access to food in their communities, which can be labelled as material deprivation and restricted choices of the inhabitants. This can be the result of large swings, and particularly increases, in food prices. A member from Pacaltsdorp reported that many people run out of money, particularly at the end of the month, and food is often the only flexible expense category.[51] If the household budget decreases, cheaper food is usually consumed. A KEF member from Thembalethu explained his experiences in the following way, "I start my month like a king – we have pap,[52] beans, and greens – sometimes even meat. At the end of the month, we often eat white bread and blue band"[53,54]. Whereas the pap meal is traditional and relatively nutrient-rich, but still not too high in costs, the white bread option marks a harsh unhealthy contrast. Eating meat often remains the exception for poor families; it is mostly low quality and residual pieces, like chicken feet (so-called walkie talkies), that are eaten. This indicated that the perceived space – the food products which could be chosen – is restricted. Food with low nutritional value often remains the only option. However, healthy and diverse nutrition is particularly important in poor areas as many are vulnerable to or affected by diseases. This was emphasised by a KEF member:

> There is so much of cancer in South African townships; it is just every form of cancer. [...] And you also know how much AIDS there is. People really need to eat very healthy when they have AIDS [...] Tuberculosis is another big disease in our area and nutrition is so important for tuberculosis too.[55]

This confronts unequal access in healthy food, which also contains a racial dimension.

Some KEF members point out that the food available has changed tremendously in quality. In different shops in the communities highly processed and unhealthy food and soda drinks have become more readily available. KEF members are sceptical about the vitamin content in fruits and vegetables that are grown from GM seeds, sprayed with chemical fertilisers, and then transported over long distances. In this regard, a member pondered,

> If it's [groceries] sitting on the shelve being processed [...] then transported to city three or four kilometres away and then it's packed there and then it's transported back again. The carbon miles are horrific. [...] The nutritional values on most of these things are zero.[56]

Today, fruits and vegetables which were previously only available in one season can be found regularly in the supermarkets. As one member explained: "South Africans are losing the connection to their food. Many don't know any more when it is time for cabbage or avocados. They just eat it year-round. Whereas I think, KEF made me aware that avos available in spring are certainly not from the region".[57] Others pointed to the antibiotics and hormones in meat raised from large-scale agriculture. Awareness about the food available in the grocery stores is shared in the group and might be termed "critical food literacy" (cf. Yamashita and Robinson 2016). Perceived restrictions in access to food and constrained choices clash with the first and third pillars of food sovereignty, which point to good quality, adequate, affordable, healthy, and culturally appropriate food and protect consumers from low quality food

(Nyéléni 2007, 1). These realities can be linked back to the prevailing commercialised food system, but nuances also reflect a clash between traditional and externally intro-duced food systems (e.g. pap, beans, and greens vs. white bread and margarine).

It was moreover indicated that in the past, food production for own consumption and even livestock raising formed a vital part of every household. However, many people perceive it as backward and inappropriate because of aesthetic issues such as odour, the responsibilities and effort entailed, or because of a lack of experience. During the neighbourhood walk in Blanco, one participant pondered, that the community was able to feed itself in the past. Today, they need to go out and buy elsewhere. This is also related to a changing retail structure, which is explained further below.

In addition, the discussions with the KEF members showed that changing lifestyles and labour conditions influence what poor people have on their plates and how food is consumed. The combination of several often-unstable jobs and precarious working conditions (e.g. long working hours and missing breaks) and long journey-times to the place of work encourage unhealthy diets. Consequently, a proper nutritional intake seems challenging.

Related to that, the private sphere of eating and food habits have changed. Members reported that in their neighbourhoods regular shared meals e.g. with the family, have become less important or family ties have been changing through migration and related job opportunities. For instance, many inhabitants of Thembalethu left older relatives in the countryside, so that the grandmother – who traditionally brought the family together and took care of regular and freshly cooked meals – was out of reach. For many it is difficult to integrate meals in a stressful day. Fresh cooking at home seems to become less doable. People eat on their way to work; children eat when they walk to school or get a snack while parents are busy with other activities. Many of the elderly people observed that younger people now struggle to prepare a meal out of fresh ingredients. In the context of low food expenditures and less social value attached to healthy food consumption, affordable and widely offered street food represents a response and drives further abstraction of eating habits. Typical street food snacks in deprived areas are of low nutritional value and high in calories, e.g. deep-fried fish and chips, pies, and sandwiches. A common South African snack is a Gatsby, a sandwich filled with French fries and ketchup and sometimes with a bit of salad.

All of this represents alienation and features of nutrition transition. Food and eating become increasingly remote and in the long run unhealthy. A KEF member summarises the nutrition problem in the following way:

> people are exposed to just one type of food [...] you can become more healthy if you eat more variety. People just eat bread and they just eat carbohydrates and little bit of meat if they can afford it and vegetables when they can afford it. And the vegetables they buy don't have much nutrients as well as being grown with chemicals.[58]

Here, the double burden of malnutrition becomes visible in a sense that individuals are consuming to much unhealthy food, which makes them obesely malnourished.

The changing nutrition landscape in George reveals strong ties to the commercialised South African food system and thus contributes to the limited choices and an increase in unhealthy diets. Although the city centre has an old town charm with long-established retail businesses, well-known national and international supermarkets and fast food chains have slowly been gaining ground. In addition, bigger shopping malls, such as the Checkers Mall and the Garden Route Mall, have opened at the fringes of the city. Supermarkets have opened in Blanco, Pacaltsdorp, and Thembalethu. A Spar store can be found in Pacaltsdorp and Blanco, and a Shoprite store is in the centre of Thembalethu Square. Nevertheless, those stores are not necessarily within walking distance for everyone, and therefore informal shops, like house and spaza shops, remain important food suppliers. Moreover, KEF members indicated that the increasing numbers of commercial supermarket outlets have changed the type and variety of food products available in the neighbourhoods, with more packaged and processed goods. Against this background, local and fresh produce is perceived as a niche market. My discussion partners pointed out that new retail stores represent huge competition for small house shops, some of which have had to close down during the last decade. Also, with the increasing commercialisation of the food system, small shops became more open to processed and packaged products. An expert of the LED describes how products from international brands are now available in every spaza shop: "Every spaza shop in here you find coke being sold, you find Nestlé products being sold [...] Those businesses are tight into the informal economy. [...] it is just cash based".[59] These brands sell their products on a cash basis in order to access the informal sphere, which they otherwise could not target. Hence, KEF members observe that the products available in these shops are increasingly dictated by large brands, promoted by professional marketing strategies and colourful packaging. Local shops risk losing out if they are not connected to these larger brands; at the same time, vegetables, fruits, eggs, or meat from local food providers have become rare in house and spaza shops.

Nevertheless, spaza shops still offer their customers some advantages which commercial stores do not. For instance, package sizes are smaller and prices sometimes lower. On the whole, KEF members see the advantages of localised consumption structures. Some of these shops also offer credit. An interviewee from Pacaltsdorp, prefers to buy his food regularly in a house shop in the neighbourhood, despite the small range of products, he explained:

> If it's the end of the month, I am a bit thin with money. Okay, so I pay later. On top of that, I get a friendly chat and if I am busy, they even bring the stuff to my house. I don't see this with the anonymous shopping temples. So, I think, we should protect these local businesses.[60]

One can see the connection here to notions of "everyday resistance", which are illuminated further in the next chapter. In his individual routine of grocery shopping, this resident avoids supermarkets and intends to support local structures. Parallels to the concept of food sovereignty can be found in this critical reflection and are explored further in the following chapter.

Concluding remarks

The overall intention of this chapter is to illuminate the perceived space and related marginalisation in everyday life. This refers to the initiative's foundation and "proposal" for change tying in with the search for alternatives and Marcuse's thoughts on the right to the city (2009) as part of the critical urban food perspective. This empirical chapter frames the initial phase of the initiative, socio-economic backgrounds, and food production in the three focus areas as well as the organisational structure. The final part sets out to engage in more detail with the initiative's evident critical reflection of the food landscape comprising members' perceived negative changes in food consumption and retail structure. Related descriptions and narratives build upon a rich body of empirical material and extensive exchange in the field.

The unique histories and locations of the three focus communities – Blanco, Pacaltsdorp, and Thembalethu – basically village-like neighbourhoods, shape the initiative's character. Apparently, KEF benefits from the rural character of its communities, a long history of agriculture in the region, and the wider ties of its members to food production, which has been part of their identity. KEF itself considers most of the members as poor; many rely partly on agriculture for their livelihoods including backyard gardeners and squatter farmers alike. They straddle land-based livelihoods and precarious employment. In this context, the theoretical concepts of the "working people", "semi-proletarianisation" and the production of space are illuminated. People do not only suffer from income poverty, but rather from diverse dimensions of poverty referring for instance to limited and insufficient access to food, health services, and education. It can be argued that food cultivation has become an integral part of social reproduction, showing parallels to Shivji's notion of the working people (2017). In response, the initiative stated its interest in promoting food security and community resilience.

The group began with the impulse of better-off community members to found a food producer initiative, which motivates people to make a difference through food gardening, knowledge sharing, and connecting with one another. Members decided to use the resources at hand – the land (e.g. fallow land on institutional grounds, empty backyards), the skills (i.e. passed from previous generations and available in the community), and the neighbourhood network. Nevertheless, the motivation to mobilise these resources and initiate change goes much deeper; it is embedded in a critical reflection on deprived realities, ranging from an increase in chronic diseases intertwined with unhealthy diets to unemployment and historical segregation. Outspoken critique against the prevailing, commercialised retail structure and the municipality's community gardening projects, to name just a few aspects, was articulated in several discussions. These conditions relate back to the context of the "right to the city", which departs from the neoliberal enclosures in modern cities and pushes people's efforts to produce space according to their needs in everyday life. Efforts to improve backyard gardens can be framed as "quiet encroachment of the ordinary" (Bayat 2000, 533). Beyond these rather quiet attempts to make a difference, widely advertised gardening demonstrations and competitions attracted the wider public. Alliances with local NGOs, the municipality, and the local newspaper were carved out. Members slowly introduced modes of exchanging resources (e.g. seeds, knowledge, tools) and strengthening community ties. Part of this is a rough organisational structure as well as diverse well-known and publicly appreciated activities. Beyond two leaders, the initiative has several area coordinators who keep the more than 700 members together and provide practical and mental support.

In addition to already ongoing activities, KEF has spatial expansion plans which include fenced-in resource sites with gardening supplies and materials in public spaces in the city. This reflects Lefebvre's notion of autogestion, which refers to citizens' interventions in producing and managing places (Purcell and Tyman 2019, 2; Lefebvre 2002, 779–780). In short, the initiative has been actively creating space for participation and contestation of the food system. In this regard, Uitermark et al. note that alternative movements (like KEF) develop "when people organize to collectively claim urban space, organize constituents, and express demands" (2012, 2546). Using the critical urban food perspective, it can be observed that related motivations and actions are examples of the inherent notions of emancipation and resistance. On the whole, KEF appropriates space according to their needs which mirrors different notions of resistance and efforts towards food sovereignty. These actions are explored and differentiated in detail in the following chapter. It is the task of the subsequent chapters to illuminate these diverse forms of resistance, alternatives, and the potential efforts towards food sovereignty in the city.

Notes

1 Interview on January 15, 2016.
2 Afrikaans word for dream.
3 Interview on January 15, 2016.
4 Interview on August 16, 2016.
5 Interview on March 6, 2016.
6 Interview on January 15, 2016.
7 Today, competitions are no longer part of KEF's work. While KEF cooperated with the municipality at the beginning, this relationship became rather weak in 2016, which is explained in Section 5.2.
8 These beds are introduced on the top of the soil; leaves, weeds or grass do not have to be removed beforehand, and the soil is not turned over (i.e. no digging). The first layer is usually cardboard or newspaper. The different layers set up on top are wet cardboard, wet newspaper, dry grass, fresh soil (which might include a bit of fresh compost, kitchen waste, and manure). These green, brown, and paper layers are put on top of each other. In this way, the bed contains organic material which will compost over time. Seedlings are planted on top. Detailed explanations of no-dig-gardening are provided by the members in the *George Herald* (e.g. 2016).
9 Interview on March 6, 2016.
10 Interview on August 16, 2016.
11 Afrikaans is one of the 11 official languages in South Africa. It has its roots in the Dutch language of the Cape settlers in 17th century, and became an individual language throughout the years.
12 Interview on March 6, 2016.
13 Interview on August 16, 2016.
14 Interview on August 16, 2016.
15 The totals do not amount to 100% as some participants did not indicate a category.
16 This is an estimation made by one of the interviewees working with the municipality, which refers to the large number of informal dwellings and continuous influx of new arrivals (Interview on March 10, 2016).
17 Interview A on August 17, 2016.
18 Interview B on March 9, 2016.

19 Interview on August 20, 2016.
20 Interview B on August 25, 2016.
21 Interview B on March 9, 2016.
22 Interview on January 15, 2016.
23 Interview on August 16, 2016.
24 Interviews in the subsequent order: January 15, 2016; March 11 (A), 2016; and August 17 (B), 2016.
25 Interview A on August 17, 2016.
26 Statistics South Africa (2011) defines shacks as "a makeshift structure not approved by a local authority and not intended as a permanent dwelling".
27 Interview C on March 9, 2016.
28 Interview B on March 11, 2016.
29 Interview on March 10, 2016.
30 Interview on March 10, 2016.
31 Interview on March 10, 2016.
32 Interview on August 19, 2016.
33 Interview on March 8, 2016.
34 Interview on January 15, 2016.
35 Interview A on August 17, 2016.
36 Interview B on August 25, 2016.
37 Interview on August 16, 2016.
38 Interview on September 3, 2016.
39 Interview on January 15, 2016.
40 Interview on August 16, 2016.
41 Interview A on August 25, 2016.
42 Interview on January 15, 2016.
43 Interview on January 15, 2016.
44 Interview A on August 17, 2016.
45 Interview on March 10, 2016.
46 Interview on March 10, 2016.
47 Interview on March 6, 2016.
48 Interview on August 24, 2016.
49 Interview on August 24, 2016.
50 Interview on August 24, 2016.
51 Interview C on August 17, 2016.
52 Pap (or Ugali, Nsima, Sadza) is a cornmeal or sometimes millet porridge.
53 Blue Band is a popular Unilever margarine brand. It is mainly made from vegetable oils, particularly palm oil.
54 Interview on September 3, 2016.
55 Interview on August 16, 2016.
56 Interview on March 6, 2016.
57 Interview A on March 9, 2016.
58 Interview on August 16, 2016.
59 Interview on March 10, 2016.
60 Interview B on March 9, 2016.

References

Bayat, A. 2000. From 'dangerous classes' to quiet rebels'. Politics of the urban subaltern in the Global South. *International Sociology* 15, No. 3, 533–557. DOI: 10.1177/026858000015003005.

Gardiner, M. 2004. Everyday utopianism. Lefebvre and his critics. *Cultural Studies* 18, No. 2–3, 228–254. DOI: 10.1080/0950238042000203048.

George Herald 2016. Medical students learn to plant own food. Published 21/04/2016. George.

George Municipality 2014. Integrated development plan 2012–2017. 2nd Review 2014/2015. George.

Harvey, D. 2008. The right to the city. *New Left Review* 53, 23–40.

Lanegran, K.; Lanegran, D. 2001. South Africa's national housing subsidy program and Apartheid's urban legacy. *Urban Geography* 22, No. 7, 671–687. DOI: 10.2747/0272-3638.22.7.671.

LED 2014. Report: Permaculture food garden pilot project. George.

LED 2016. Project plan: Household food security project phase 1: March – June 2016. George: George Municipality.

Lefebvre, H. 1991 [1947]. Critique of everyday life Volume I: Introduction. London: Verso.

Lefebvre, H. 2002. Comments on a new state form. *Antipode* 33, No. 5, 769–782. DOI: 10.1111/1467-8330.00216.

Marcuse, P. 2009. From critical urban theory to the right to the city. *City: Analysis of Urban Trends, Culture, Theory, Policy, Action* 13, No. 2–3, 185–196. DOI: 10.1080/13604810902982177.

Nyéléni 2007. Synthesis Report Nyéléni 2007 Forum for food sovereignty. Forum for Food Sovereignty. nyeleni.org/IMG/pdf/31Mar2007NyeleniSynthesisReport-en.pdf, Accessed on 09/01/21.

O'Brien, K.J. 1996. Rightful resistance. *World Pol.* 49, No. 1, 31–55.

Purcell, M.; Tyman, S.K. 2019. Cultivating food as a right to the city. In Tornaghi, C.; Certomà, C. (Eds.) *Urban gardening as politics*. Oxon: Routledge, 46–65.

SACN 2014. George land of milk and honey? *SACN Research Report*, University of the Free State. http://www.sacities.net/wp-content/uploads/2015/10/George-report-final-author-tc-3.pdf, Accessed on 20/03/2020.

Shivji, I.G. 2017. The concept of "working people". *Agrarian South: Journal of Political Economy* 6, No. 1, 1–13. DOI: 10.1177/2277976017721318.

Siebert, A. 2020. Transforming urban food systems in South Africa: Unfolding food sovereignty in the city, *The Journal of Peasant Studies* 47, No. 2, 401–419, DOI: 10.1080/03066150.2018.1543275.

Statistics South Africa 2011. Census 2011. Population dynamics in South Africa. Report No. 03-01-67. Pretoria. http://www.statssa.gov.za/publications/Report-03-01-67/Report-03-01-672011.pdf, Accessed on 16/09/20.

Statistics South Africa 2018. General household survey 2018. Pretoria. http://www.statssa.gov.za/publications/P0318/P03182018.pdf, Accessed on 19/06/19.

Uitermark, J.; Nicholls, W.; Loopmans, M. 2012. Cities and social movements. Theorizing beyond the right to the city. *Environment and Planning* 44, 2546–2554. DOI: 10.1068/a44301.

Western Cape Government Provincial Treasury 2015. Socio-economic profile George Municipality. Working Paper. Cape Town. www.westerncape. gov.za/assets/departments/treasury/Documents/Socio-economic-profiles/2016/municipality/Eden-District/wc044_george_2015_sep-lg_profile.pdf, Accessed on 05/09/20.

Yamashita, L.; Robinson, D. 2016. Making visible the people who feed us. Educating for critical food literacy through multicultural texts. *Journal of Agriculture, Food Systems, and Community Development*, 269–281. DOI: 10.5304/jafscd.2016.062.011.

Zhan, S.; Scully, B. 2018. From South Africa to China. Land, migrant labor and the semi-proletarian thesis revisited. *The Journal of Peasant Studies* 45, No. 5–6, 1018–1038. DOI: 10.1080/03066150.2018.1474458.

4 Proposing food sovereignty

The previous chapter provides first details into how the urban agriculture initiative KEF is exposing inequalities and zooms in on the origins and mobilisation of the group. Using the framework of the critical urban food perspective, this chapter adds further details and focuses on the step of "proposing" alternatives. Particularly the production of space and everyday forms of resistance are key to understand the initiative's critical reflection of the prevailing food and agriculture landscape. The chapter is divided into five sub-sections which use the food sovereignty pillars as guideposts and separately introduce day-to-day practices, challenges, and ways of resisting. Some of the findings exceed the parameters of the food sovereignty pillars, other parts are rather narrow. Consequently, only five clusters are framed here. While the people of the initiative are the focus, several interventions are contrasted with the broader political landscape and actors around KEF. For instance, the George Municipality and government departments both support and restrict the initiative's practices.

Following Scott's work, different forms of resistance can be traced back to material needs including food, work, land, or charity (1985, 289). Everyday practices of the group have innate aims; the sharing of permaculture knowledge, for instance, seeks social and ecological change. In fact, KEF has been fighting on different fronts, dismantling inequalities and constructing alternatives. On the whole, the initiative presents overt and covert forms of everyday resistance. Before providing case study detail, relevant conceptual considerations as part of the critical urban food perspective are outlined.

Several scholars in the field of "critical agrarian studies", including Scott (1985) and Kerkvliet (2009), speculate that peasants are the most revolutionary class; they have agency, and most of their resistance to oppression is carried out in forms not legible to outsiders (e.g. unstructured and covert). This kind of everyday politics does not necessarily imply that people are organised politically (Kerkvliet 2009, 229). While Scott's work is premised on patron–client relations prevalent in non-capitalist agrarian settings, scholars like Bayat and Roy observe related "quiet rebels" and "quiet encroachment" in informal areas amongst marginalised dwellers (Bayat 2000, 536; Roy 2011, 228). Examples include electricity or grocery theft or using paid labour time for regenerative activities. This follows Scott's take which refers to the small-scale, almost invisible activities – those which are unlikely to attract the attention of the powerful and thus carry the risk of punishment. Such

DOI: 10.4324/9781003182634-4

resistance is "covert" because of the "individual contention" which Scott was keen to understand. Scott further distinguishes between different forms of covert acts of resistance, referring for instance to pre-planned and incidental acts (1985, 292). This is what can be taken from Scott and cuts across agrarian space. From the broad perspective of urban studies, Roy emphasises that exploring covert resistance helps to unveil "forms of popular agency that often remain invisible and neglected in the archives and annals of urban theory" (2011, 224).

Resisters may moreover seek equality and implementation of existing rights and rule of law (O'Brien 1996, 34). Resisters highlight the gaps between the promised and the actual enforcement in an opportunistic unpacking of these shortcomings (O'Brien and Li 2006, 38). While the right to the city suggests broadening the horizon of the possible and thus moving beyond existing rights, O'Brien's concept of rightful resistance refers to important intermediary steps. In sum, these considerations imply enforcing existing rights and values or even granting new rights. The concept of rightful resistance is rooted in peasant struggles in rural China. O'Brien illuminates his thoughts with cases of organised villagers rejecting elections through occupation of township offices and demanding fair procedures after observing several incidences of corruption (1996, 38). The concept can be applied across a range of situations of complaint related to disregard of official values, making it useful in analysing struggles of urban food producers. In contrast to Scott's covert resistance, O'Brien's notion goes beyond individual and rather covert struggles: "It is a product of collectivities living in a specific political economy and socio-cultural setting: of people with histories and moral understandings who are grappling (individually and together) with […] neoliberalism and many other challenges" (2013, 1053). Nevertheless, O'Brien admits that there are several parallels to covert everyday resisters as both forms of resistance often lack organisational resources and collective consciousness (1996, 34).

Bayat also points to overt resistance initiated by the "political poor" and frames some parallels to O'Brien's rightful resisters. In this regards, he explains that "community associations […], consumer organizations, soup kitchens, squatter support groups, church activities and the like were understood as manifesting organized and territorially based movements of the poor who strive for 'social transformation' and thus their […] share in urban services" (2000, 540). It is the lack of adequate access to certain resources and public services such as water, electricity, or food, and related rights as well as other officially promoted values, that propels these organised groups' claims and struggles.

Building from these considerations on resistance in the context of the critical urban food perspective, the subsequent sections of this chapter provide rich case study insights using the food sovereignty pillars as an overall frame. A diverse range of empirical material feeds into this analysis.

Local access to nutritious food

This section is framed by the first and third food sovereignty pillar. The first pillar of food sovereignty in the Declaration of Nyéléni highlights "the right to sufficient, healthy and culturally appropriate food for all individuals" (Nyéléni 2007, 1). It moreover emphasises that food is not another commodity for agri-businesses. However, as elucidated in Chapter 2, South Africa is facing rapidly transforming dietary patterns comprising mounting health concerns as well as a triple burden of malnutrition. The prevailing retail structures do not directly meet the needs of the poor. In general, advancement of governmental food policies particularly regarding nutrition, for instance, subsidies for healthy food products, regulations on food labels, and related food advertising, remain future tasks (Claasen, van der Hoeven, and Covic 2016, 3). To be sure, these are only some of the many aspects which need to be addressed to bring the idea of "food for the people" to fruition. Pillar 3 goes further and highlights the role of a localised food system, in which providers and consumers work together and thus are "at the centre of decision-making on food issues" (Nyéléni 2007, 1). In this way, food sovereignty is concerned with the protection of "food providers from the dumping of food and food aid in local markets" and consumers "from poor quality and unhealthy food" (ibid.).

Much of KEF's motivation and its practical work show parallels to these principles. While there is a strong interplay between the pillars in terms of demanding appropriate food and food systems, the third pillar frames this with a kind of protectionism of local food systems including food consumers and providers. KEF members represent both consumers and providers. Hence, concerns and needs from both sides are enshrined in the lived practices.

The previous chapter already provided important insights into the perceived nutrition landscape and, directly related to that, members' felt exclusions and the ways in which food consumption are abstracted in the prevailing food system. This part illuminates the lived space, and thus the alternatives which the initiative creates through food production and consumption. In general, the demand for healthy and appropriate food is evident in the initiative's work. One of the founders of KEF summed it up as follows, "Kos en Fynbos stands for food security and biodiversity, they complement each other and that's what community resilience is built of".[1] Newspaper articles in the *George Herald* often introduce KEF as a "food security initiative" (e.g. 2014a). While this is part of the lived practices and motivation, it also points to the challenges for members in their daily lives. Hunger, unhealthy diets and the wish to be more independent (resilient) from the commercialised food system are driving the interventions.

A mobile-based consumer survey[2] of the Local Economic Development Unit (LED) of the George Municipality yielded important information on where people in disadvantaged areas make their purchases including fresh fruits and vegetables, meat, and groceries. Results reveal that more than 50% of food products are purchased in the area (with very high numbers for vegetables and fruits) in Blanco; numbers are lower in Thembalethu (about 40%) and Pacaltsdorp (about 12%) (LED 2013). The informal sector was represented in the survey by the categories "house/spaza shops" and other places like food stalls. People in Thembalethu could also tick the box "Sandkraal Road", which is the main road in the area and hosts a wide range of food stalls and informal shops. A key finding is that the informal sector is still a considerable player in feeding these communities. In the survey, the municipality defined the spaza and house shops as opposite to formal supermarkets. An expert of the LED defined spaza shops as part of the informal sector and thus difficult to control; and, importantly, they are not paying tax.[3] In general, several studies have revealed the distinct role of the informal sector in food provision in poor areas in South Africa (e.g. Peyton, Moseley, and Battersby 2015; Skinner and Haysom 2017).

There are different reasons for the variation of these numbers across the three areas. For instance, the number of formal shops and supermarkets in Blanco is low, which is not the case in Pacaltsdorp. In Thembalethu, the percentages for informal purchases might actually be higher as several places in the area, like the Thembalethu Square, comprise both informal and formal retail structures. The square hosts a big Shoprite supermarket but at the same time the taxi rank has long been a perfect informal selling spot for vegetable and fruit traders as well as for snacks, meeting needs not targeted by the supermarket structures. The vendors are directly in front of people waiting for buses. Some of them are selling local herbs and products harvested from gardens and farms. The social worker in Thembalethu mentioned that they even take care of special orders: "We like green peppers on Sundays in our food […] Sometimes I buy 7 kilogrammes to bring here [to the Thusong Centre]".[4] This kind of customer service is not available in supermarkets. Small market operators remain the easiest accessible salesmen for those growing food in the city in all three neighbourhoods.

The message from the briefly presented data and the issues raised by the initiative show that the formal retail system is not necessarily the primary food access point for everyone, which is the assumption of many policymakers and planners. Throughout the neighbourhood walks, it was clearly visible and highlighted by participants that the informal sector plays an important role in selling local gardeners' and farmers' produce (see Figure 4.1). In the LED survey, spaza and house shops are considered as opposite to formal supermarkets. In discussions experts of the LED defined spaza shops as part of the informal sector and thus difficult to control; and, importantly, they are not paying tax. Similar findings regarding the informal sectors role are provided by Peyton, Moseley, and Battersby (2015), who highlight that the informal economy still plays a crucial role in food access in poor areas in Cape Town, and that the commercialisation of the food sector contributes to a higher prevalence of food insecurity in poor neighbourhoods.

This tendency to purchase food products from local shops is of course not only related to the idea of strengthening the local economy; there are various reasons behind it. First, people's budgets are tight; if the journey to work does not take them to an area with a supermarket, people buy what is available within walking distance as they simply cannot afford additional trips for food purchases. Second, informal structures offer several advantages as indicated above, e.g. smaller package sizes, credit, the opportunity to order specific products, and sometimes even delivery. It might be interesting to dig deeper into the reasons for different purchase behaviours. However, a more detailed exploration would exceed the scope of this project. In addition, these consumption and food patterns, as well as related challenges such as time and financial constraints, illustrate how difficult it is for individuals to make their own choices. They represent obstacles for food sovereignty's first pillar regarding "A world where all peoples […] are able to determine their own food-producing systems and policies" (Nyéléni 2007, 1). At the same time, the lived experiences show that decision-making in the food retail landscape is not directly targeted at a local level (pillar 3). In general, supermarkets and spaza shops have the freedom to choose what kind of food they offer; there are limited restrictions from the official side (i.e. food governance). Hence, the protection of consumers from low-quality, unhealthy food and food tainted with GMOs (according to pillar 3) remains challenging. However, the interventions of organisations like KEF propose local alternatives and thus carve out options that run parallel to those food sovereignty pillars.

The following parts focus on the alternatives created by the initiative. On the whole, KEF members value informal structures and want to see a burgeoning localised economy, which also provides a space to sell their produce. Hence, the initiative is rather critical about the growing supermarket structures: "we don't need that Pick'n'Pay, Shoprite stuff. It is time for the government to learn [...] we don't want that".[5] Localised food production and consumption are manifested in different ways in the work of the initiative and are part of the lived space, which comprises a variety of alternatives to an abstract and remote food system.

In one of the interviews, a KEF member described the overall motivation of the initiative as "a vision where everyone in town has a food garden, and no one wants to buy food at the supermarket anymore and we have an anti-consumerist revolution. And we are all hippies and we are all happy. Which is also not completely realistic".[6] While this certainly goes beyond the feasible, it nevertheless shows the horizon that KEF envisions. This kind of thinking implicitly expresses the initiative's intention to go beyond the capitalist (food) system, which – according to Lefebvre – can never fully colonise everyday life (1971). Hence, mundane practices of growing and sharing food as well as utilising local networks in terms of access and consumption are not only a self-help strategy but also a kind of resistance to the prevailing food system, which restricts people in their choices and access. This relates with the critical urban food perspective, specifically Bayat's notion of quiet encroachment (2000).

A community worker and member of KEF in Thembalethu admitted that many are struggling with hunger and health issues in the neighbourhood; but was also convinced that "if you plant for yourselves in your backyard you gonna eat regularly".[7] Although regular availability of food and related harvests depend on many factors including the size of the garden and skills, this message emphasises the window of opportunity the initiative creates. One of the leaders goes even further and refers to food purchases as the only flexible expenditure category:

> People are poor because they have no money. [...] No money to buy food. So, food is the primary cause of poverty. Then let's have everyone just have their own garden. Cause then they don't need money, they got food. That's very important to associate the fact, that they don't really need money to have food.[8]

Although this argument is oversimplified, both interviewees reveal the importance and opportunities enshrined in household food production. While the prices, quality, and availability of products are matters of concern for members of the initiative, self-provisioning creates a (partial) independence from the market and returns the control over food and nutrition to the people. One coordinator from Blanco summarised this as follows: "Use your own veggies and your own herbs. So, we don't depend on other people. We don't need the doctor or anyone".[9] Based on her experience with KEF, another KEF member from Thembalethu strongly recommended to the poor in her neighbourhood: "they must sustain themselves and they must eat veg. It's not the supermarket which is taking care. Because until now the sickness and chronics is too much".[10] The health concern in the communities is an important driving force of the initiative. There is a common understanding among the group that the existing formalised food structures do not meet the need for fresh and healthy produce. KEF tries to promote its message by showing the little effort it takes to garden and to change certain consumption patterns: "Our bodies are so incredible; it just takes a bit of nutrition and a bit of fresh stuff to actually change your whole constitution, your whole outlook on life and personal health".[11]

Following these thoughts, own food production is certainly of high priority for the initiative. Therefore, available material space[12] (beyond its restrictions) comprising land, soil, water, air, crops is extremely important. Departing from this material space, members carve out new representations of space that build on a critical examination of food consumption practices and health. In this way, they uncover unjust developments in the city (e.g. restricted food access and choices). Their lived space goes beyond the restricted availability of healthy food, isolated production and consumption practices. As part of this strong community engagement, KEF initiators felt that the group shaped a new kind of appreciation and value in food production:

> people who said before 'I am poor, I have to grow my own vegetables' [...] we were saying when we were going around, 'no, no, no you are the future [...] the feed stores for the future — the clean feed. [...] you are going to be able to teach the next generation how to do it.' So suddenly this people had a pride.[13]

Besides food production per se, alternative sharing networks (e.g. for food, seeds, seedlings) play an important role in the initiative. The local doctor explained: "They eat what they can, and they dish out to their neighbours. It is very much a sharing community vibe, and we wanted to promote that".[14] Members even prepare food together and at the same time try to contribute to nutritional quality. For instance, in the gardening demonstrations: "when it came to taking a break from doing the course, we brought bananas, baby tomatoes – we didn't bring chips and biscuits. I didn't bring rubbish cool drinks. We bring water".[15] Again, these lived realities represent encroachment of the ordinary in relying on healthy food and fresh food preparation in the community. Moreover, these practices are not hidden; they are visible for other community members and even outsiders. Another example is the Khoisan breakfast organised by the group during the national science week in 2015. These practices also promote informal sources of food, the sphere of informality. Such interventions and collective actions bring the ideas of the members to fruition, shape people's actual experience of space, and are integral to the lived space; they shape the material space and create a new knowledge and sense about space (conceived space).

KEF's conceived space comprises self-sufficiency, localised control, and a common space for growing food.[16] Their critical reflection on the commercialised food system values local alternatives of which spaza shops and food stalls could be important elements. These also represent some of the limited opportunities for members to sell their produce. While members are aware that it is impossible to abstain from buying basic products, they are interested in reducing the number of things to be bought. This would also reduce the need for money. The coordinator from Blanco explains that, "you can sell your veggies and then you only buy the things you want like sugar and fish oil [...]. You buy things you need that you can't grow".[17] Moreover, the idea of self-sufficiency can only be realised if certain resources are available, for instance the labour to work the soil and the land to grow food. In spite of the limits to these resources, food producers are creating spaces of self-reliance under capitalism. But those are never entirely free of the external conditions introduced by the food system: it is important to highlight that gardens are not the only source of food and at least a few ingredients are bought outside. At least partial connections to – and thus acceptance of – the corporate food system remain. These limitations are further discussed in Chapter 5.

The different activities undertaken and opinions expressed can be termed conscious political action. These everyday forms of subsistence contribute to self-determination. In the long run, the lived space (spaces of representation) refers to a higher quality and diversity in food.

Valuing food providers

Food sovereignty's second pillar focuses on those "who cultivate, grow, harvest and process food; and rejects those policies, actions and programmes that undervalue them, threaten their livelihoods and eliminate them" (Nyéléni 2007, 1). As outlined earlier, the South African government's attempts to expand the smallholder sector and facilitate land access are far from successful. In addition, urban food system governance needs to be responsive to the needs of small producers and poor consumers.

This section frames the government's role in facilitating food production activities and the interplay with KEF. There is a Janus-faced aspect to governmental interventions. On the one side, the municipality and government departments appreciate the ongoing gardening activities and incorporate them into their own projects. On the other side, there is evidence of governmental neglect of the initiative and its members. Both these facets are explored in this study. In general, the local government is over-burdened with the fallout of high unemployment rates, fails to adequately support self-help strategies, and to introduce public policies facilitating healthy food provision beyond the power of the market. This kind of critical observation is part of the book's analytical perspective which simultaneously also uncovers alternatives.

In addition to marginalised groups' difficulties in accessing sufficient food, many of the KEF members struggle to make a living: in the absence of stable employment opportunities, they rely heavily on grants, or on a combination of different jobs. Against this background, it is possible to glimpse identities such as that of the (often informal) labourer and that of the food producer (mostly for social reproduction). For instance, many of the members in Thembalethu have roots in the Eastern Cape. The pathway from the countryside to the city – from the land to wage labour and back to the land – is summarised by a farmer: "I left the Eastern Cape about 20 years ago. [...] working here and there [...] and finally ended up in Thembalethu. First, I worked as a construction worker, but perspectives were limited [...]. So, we decided to work this land".[18] These realities are ignored in two ways by the government, which fails both to integrate these marginalised urban dwellers sufficiently in the (formal) labour market and to improve conditions for urban food production. Hence, the initiative's activities offer a double self-help strategy: to provide food and to earn a bit of money. The group exposes these failures on the part of the government and shows that its members' reliance on farming is also not fully recognised by the authorities, which fail to provide land access and adequate support, for instance in terms of market access to food producers.

Maintaining a long-term vision for urban food production is not part of the government programmes in place. An expert of the Farmer Support and Development Unit (Department of Agriculture), Western Cape Government (funded by the National Department of Agriculture) clearly stated in the context of the implemented programmes that "land is a challenge in the garden route area. So, it's not easy. The communal lands are earmarked for residential. So, they can't have long-term leases".[19] This limits the support opportunities of this unit as well as the LED of the George Municipality. The demand for land and governmental responses are further analysed in the following chapter.

Government interventions in terms of farming and gardening in the city focus on food security and in the long run also on income generation. In the 2016 project plan of the LED, interventions were also termed "quick win projects", which already indicated the short time horizon of the unit (LED 2016). Over the last decade, marginalised communities have been targeted with two kinds of interventions by two different government units, which overlap in their mission. First, the Farmer Support and Development Unit supports smallholders and gardeners in running a garden or a small farm in urban areas. The second and more dominant intervention in KEF's everyday practices was introduced by the LED and comprised several time-limited gardening projects. The work and interventions of both government units are sketched out below.

Farmer Support and Development Unit

The Farmer Support and Development Unit offers assistance to potential gardeners upon application. An expert of the Farmer Support and Development Unit describes their interventions and various funding avenues in urban and rural areas in the following way:

> we would fund gardens, household gardens from 4,500 Rand per garden to a smallholder or commercial setup that we provide infrastructure support. So, most of our funding goes into infrastructure support. We don't give money to a client.[20]

A service provide is responsible for the respective implementation. Funding for a gardens basically comprises tools, seeds, and seedlings. Overall, projects of the unit are strongly determined by the national government. The expert elucidated:

we have benchmarks per quarter how many households we have to service, how many food security projects we have to service, and within our food security we fund up to 120,000 Rand. That is small gardens, bigger gardens, tunnels.[21]

Besides the intention of promoting food security, the unit is interested in generating surplus produce which can be sold. However, creating and sustaining these market linkages remains one of the prospective interventions.

The decision on whom to support takes place in close cooperation with the ward counsellors. Those who are supported are "the poorest of the poor".[22] Besides the implementation infrastructure, beneficiaries receive training. In discussions, it became evident that several garden implementations were not successful, as people were not dedicated to growing their own food. The director considered KEF helpful in following up on these gardens. If there is more capacity and the gardens are running well, further support is possible, including an upgrade to a smallholder status, but long-term land access is an essential criterion for further support.

One of the KEF members who had previously worked with the Department of Agriculture reflected:

> If they can say they helped 300 small-scale farmers, they just want to do that. Whereas looking a bit deeper is missing, how they helped, and how sustainable are those 300 farmers. Are they still continuing after 5 years [...] the Department of Agriculture is [...] not functioning very well, and it makes me quite sad.[23]

In this sense, KEF generally considers top-down projects, which fail to spur long-term commitment, to be insufficient. This critique was also reflected by the representative of the Thembalethu farmers' union who has had experiences with the Department of Agriculture's projects. In his words:

> Follow up is missing if the government implemented something especially for this food production. [...] one document said they fight poverty and lack of jobs and people were getting treatment with food. [...] they don't do a follow-up. Monitoring whether it helped or not. And people like to work silos.[24]

"Silos" refers to the representatives of different governmental units and often unclear responsibilities across different municipal and governmental units. For those on the ground, it seemed difficult to refer back to those who are in charge of the interventions, which is also related to the role of service providers in implementation.

Local Economic Development Unit

The LED has been directly cooperating with KEF, but has nevertheless been the subject of some criticism by the initiative over its preference for "big number" projects, its lack of a long-term vision, and the absence of appropriate monitoring. For instance, a member from Blanco pointed out that, "We [KEF] don't promise you something, and then we are never coming back. Everyday, they must eat, everyday they must look after themselves, and the municipality does it once a year. But the people look after themselves everyday".[25] The problem also becomes evident from an interview with an expert of the Community Markets and Food Gardens Project (LED):

> from now I am thinking of having about 200–250 gardens. And in my second year I am thinking of 500–700 gardens. And in my third year we will probably be over a 1,000 gardens. [...] they will be all sustainable. [...] I am not going back to the others. Just continue, continue. So, you will just go to the others.[26]

Quantity plays a pivotal role in these interventions. Moreover, the LED's gardening projects only have a short project duration of three to four months, and follow-up is never guaranteed as it depends on the municipal budget, the constellation of political parties and persons in charge at the municipality.

The LED's interest in KEF was aroused in 2013 after the initiative's first successes in reaching out to many people in Blanco. In contrast, the LED was struggling to achieve a long-term commitment in its pilot gardening project in Touwsranten and Wilderness Heights (LED 2016, 4). Under this project, gardening material and training were provided through an appointed service provider, which was completely different to the community-led project approach of KEF.

After supporting KEF's first competition in 2013, the LED pushed for further competitions, invited KEF to initiate gardening demonstrations in Touwsranten (where the LED project had previously failed) and to conduct outreach in other areas in George. In this way, the LED tried to guide and structure the initiative. The LED, in cooperation with KEF, assigned coordinators, remunerated them with a small allowance, and expected regular reporting on ongoing activities. This rapidly initiated cooperation helped KEF to grow and to reach out, but it also challenged the initiative with additional administrative workload. In this regard, one of the founders explained,

> they don't have any other income. It's a very small allowance but it does allow them to have petrol money, to be able to go somewhere to show how a garden is made. [...] All of a sudden, the coordinators need to give feedback reports. These are not the kind of people who are going to give feedback reports.[27]

It became evident that frictions were created.

Registration forms for competitions were part of that administration. In this context, the person in charge in the LED criticised some of the coordinators, claiming that they were providing incorrect numbers of participants, which were often higher than the real number.[28] The LED also set criteria for the competitions and participants and decided what prizes would be appropriate. Thus, the involvement of the municipality also created tensions. One of the members described the concerns of one founder:

> He was just worried that the municipality would hijack the project and then kill it. Which is what happens to a large degree with a lot of projects [...] on the other hand [the LED] felt that we were starving the project unnecessarily from great resources that would allow us to grow even further.[29]

In this context, the initiative was mainly concerned about a so-called handout culture, which implies that practical, hands-on work is getting. Instead, leaflets are provided.

According to several members, the loosely organised initiative was put under more and more pressure to fulfil the criteria introduced by the LED. The LED wanted to see clearly defined leadership which could be addressed in case of questions, whereas the initiative had originally wanted to avoid hierarchical structures. In a meeting in early 2016, the group bowed to pressure and elected two leaders (see Chapter 3). However, it also took the chance to reaffirm its values as a groundswell initiative: "We wanted to make sure that we remained true to the movement" (George Herald 2016). In late 2015 and early 2016, KEF reported that it found the LED involvement and demands increasingly intrusive and was even afraid of losing its grassroots character. Several members preferred to part ways with the municipality. One member summarised what they were thinking: "there is too much organisation and too little action".[30]

In sum, the efforts of the local government in the broader field of urban agriculture seemed to undermine the ideals of the initiative and lacked sensitivity concerning the situation in the communities. The local doctor shared another experience: "They [the municipality] planted trees with government subsidy [...] and the community just came and snapped them of".[31] The interventions were planned without the dwellers and thus caused frictions. These experiences call to mind the second food sovereignty pillar which "rejects those policies, actions, and programs that undervalue them [food providers]" (Nyéléni 2007, 1).

In response, the initiative introduced membership certificates, which were to be put outside the participating houses, and reaffirmed its groundswell character in a meeting. The idea was to strengthen the perception of the group as a grassroots movement and to gain more attention for members' specific needs: "Kos en Fynbos started from nothing. And it grows and grows bigger until it has the right people. We can do it alone. It takes a will to raise a child. [...] If we need something, the municipality must be able to help us and not to control us. [...] Because this is Kos en Fynbos not Kos en Municipality".[32]

Despite the disappointing experience, certain elements of the cooperation with the LED remained and became important components of KEF's daily work. For instance, area coordinators, which were previously remunerated by the municipality, redefined their role beyond their previous responsibilities, continued to carry out their commitments, and developed future perspectives. One member explained the area coordinators continued vision in the following way:

> Kos en Fynbos is a vehicle for community integration for them to get to know their community to establish systems of sharing, of trading and also linking the trading to town to businesses and for them to access markets and to have a bit of an income. [...] So, the reason why the area coordinators stay involved is they see it as movement of hope.[33]

In this way, coordinators have acted as important information hubs interlinking the communities and building potential links to markets.

Unemployment and market access: the role of the informal sector

In a newspaper article, the mayor of George indicated that through the work of KEF, "The community is taught to help themselves" (George Herald 2015). In this context, it can be argued that the government has shifted responsibility for food provision and income-generating opportunities onto individuals. What is missing, as one of the members suggested, "is digging a bit deeper".[34] This has two elements: on the one hand, there is a lack of understanding of the lived experiences of the target communities and a failure to critically evaluate the implemented interventions. On the other hand, there is an absence of long-term opportunities, like official land access for those in need, official employment opportunities, or market access for those who are successfully running their gardens and are in the position to sell part of their produce.

In fact, access to markets and income generation opportunities are mentioned in the programmes of the LED and Farmer Support and Development Unit. Thus, one project component of the LED focuses on the introduction of local community markets. This was still in its initial phase in 2016, and far from being a spot for small urban producers to sell their harvest. Rather, entertainment and commercialised food stalls are the reality. An expert of the Farmer Support and Development Unit added another perspective. It was pointed out that some beneficiaries of the programme have market connections to small shops in town, e.g. selling honey to a coffee shop in town. However, the unit does not actively initiate those connections, and thus, the unit was considering including this kind of networking in future plans. Many urban food producers thus still seek an entry point to such rather exclusive networks and connections to potential purchasers.

Another project component of the LED has been targeting the huge problem of unemployment in George. The overall unemployment rate in the municipality was about 15.3% in 2016, but it was much higher in the focus communities (Western Cape Government Provincial Treasury 2018, 367) (see Chapter 3). The Worker's Collection Point (previously called "Men on the side of the road") is a meeting point for those seeking short-term labourers (mostly day-based) and those who are available for work, such as construction or maintenance work. On the compound of the Worker's Collection Point there is a garden, which workers maintain while waiting for a job opportunity. According to those managing the site, it is also a training ground to improve gardening skills. In addition, the Worker's Collection Point provides gardening advice and seedlings. Another plan of the LED and the Farmer Support and Development Unit was the implementation of an AgriPark, a food packaging, and processing factory, to create jobs and to use the local produce of gardeners and farmers. However, this idea did not materialise.

For market and employment opportunities, the informal sector often remains the only option in the marginalised areas. However, it seems to be a red flag in governmental planning and lacks recognition. An expert of the LED admitted:

> as government I think we are completely out of touch with this kind of informal economy. We don't have a clue what is happening there. Often because us, the majority policy-makers, don't live in those economies. [...] policy-makers in general don't really understand the realities of people.[35]

It was moreover highlighted that people in marginalised areas quickly move out in a middle-income area as soon as they have a better job. While the government side seems to be out of its depth when it comes to the informal sector and structures, KEF is already ahead. Informal labour and grocery markets have been a vital part of the group's lived space and provide an important alternative. The often hidden and informal ties to markets and sharing networks are a vital component of KEF, and can be seen as an everyday economy and social protection net (e.g. food provision) from below. These notions chime with Shivji's working people, who suffer precarious integration in the labour market and thus engage in diverse activities to secure social reproduction. The initiative's linkages to the informal sector can be broadly clustered in five overlapping components: selling points, sharing and exchange networks, resource sites, land demands, and job creation.

First, the rather small group of squatter farmers and a few backyard gardeners in the group have selling ties to spaza shops and food stalls (informal) as well as restaurants and cafés in the city centre (formal). Some food producers know the local shops and also exchange contacts with other members. They have informal agreements with shops in town; for instance a farmer from Thembalehtu regularly sells spinach to a coffee shop. One member explained about the farmers in Thembalethu, "They sell it wherever they can. Sometimes the supermarkets or they go out and load up. Other times they do both. People come from the spaza shops to buy the cabbages. Or the people come from town and buy".[36] The social worker from Thembalethu describes how easy it is for the food growers to access informal markets like the one on the taxi rank: "You can also be entrepreneurs. You can also have your own business. If you plant green peppers. We like green peppers [...]. You just stand there in the corner or you go to town. You can buy or can exchange".[37]

Second, these informal networks also include exchange and sharing in the communities beyond the monetary component. Many prefer to share their surplus or exchange, including seeds and food. The regular meetings are a vital opportunity for sharing.

Third, the initiative introduced resource sites and, at the time of the research, was in the process of setting up more including tool libraries. For instance, one of the members in Pacaltsdorp has a community resource site on her compound. She collects manure, dry grass, cardboard, and wire. Members come and collect and bring material. Although these sites are on private grounds, the spots are of collective value for the community. While the exchange of tools already takes place, tool libraries and further resource sites including waste management are on the agenda. Thus far, the municipality has not been able to make land available or to open sites on public grounds (e.g. the Workers Collection Point).

Fourth, space for farming and animal raising is restricted. Hence, food producers have diverse ways of claiming material space, e.g. occupying government land at the fringes of Themablethu, allowing cattle to roam around freely, and growing food on public institutional ground (e.g. schools). Especially in Thembalethu, many animals are not registered, thus it is often difficult for the officials to know whose livestock it is. This relates to Scott's everyday forms of resistance (1985), which try to avoid public attention and remain rather hidden. In another example from Pacaltsdorp, some neighbours are maintaining a big garden in a crèche. Part of the harvest is used by the crèche, the major share goes to the growers. The land problem is discussed further in the following chapter.

Finally, some members are engaged in the informal labour market. Discussions in Thembalethu, especially, revealed that some earn a bit of money in the local shops in the neighbourhood or do the cleaning in some shops or private homes in town.

The critical urban food perspective reveals these informal structures and networks parallel with Scott's everyday forms of resistance (1985) and Bayat's notion of quiet rebels (2000). Hence, the engagement in structures which are not necessarily beneficial to the formal economy and in several cases remain invisible for outsiders, apparently even the government. As highlighted in the quote from the LED manager of the municipality, who felt out of touch with these kinds of so-called shadow economies. Bayat goes further in his assumptions, suggesting that manifestations of this informal encroachment might be overt and therefore add a political component to the lived realities of informal farmers, workers, or hawkers, the so-called political poor (2000, 540). Thus, ties to the informal economy might even be interpreted as a reminder of the need and demand for more appreciation or integration of economies from below into the formalised economy.

Informal structures point to social capital and the creation of alternatives to the restricted and exclusionary formal economy as well as to the lack of social protection. Only by considering and valuing these structures can the government advance any form of meaningful support and engagement with them. Certainly, it can be difficult to distinguish between the formal commercial and the informal – this was also the case for my interviewees. For instance, selling produce for cash to a coffee shop was considered a formal purchase structure. Having a closer look at peoples' aims and intentions reveals that formal and commercialised structures are not always the desired end. Informal selling structures offer some advantages, such as being easier to access and more flexible.

Access to land

The fourth pillar of food sovereignty focuses on local control "over territory, land, grazing, water, seeds, livestock and fish populations" of local food providers, it "advances the right of local communities to inhabit and use their territories" and "it promotes positive interaction between food providers in different regions" (Nyéléni 2007, 1). In this context, the socially oriented and environmentally friendly use and sharing of these resources is highlighted. Borras et al. point to the urgent need of many groups for land and propose "democratic land control" for food sovereignty (Borras, Franco, and Suárez 2015, 606). This pillar emphasises democratic control in the agri-food system and the participation of civil society in decision-making processes regarding production, distribution, and processing of food. Closely linked to the right to the city, urban food providers need space of representation and participation in local governmental policy-making circles. As discussed above, knowledge, tool, and seed sharing are integral parts of KEF's work. In this way, KEF has been contributing to the localisation of control over resources required for food production. There are moreover cases in which KEF food producers share pieces of land (see previous chapter). However, beyond these vital sharing activities, land access remains one of the essential demands. Although many members grow food in their backyards and in some areas on public land, sufficient and secure access to land remains challenging, particularly in densely populated neighbourhoods. One of the founding members indicated that, "land access is a big issue [...]. We really need a long-term way to make land available because then it will help the people not only to feed themselves and the family but also to earn a little bit of income [...]. We really need this independence".[38]

In this way, the material space (spatial practice) of the group is limited. Although local government bodies have introduced efforts to improve food security particularly through urban gardening (as noted above) and several interventions in fighting land inequality have been made in the aftermath of the apartheid era, urban land demands remain neglected. Being affected by these inequalities, KEF has been active in exposing weak government interventions and the lack of consideration for land demands on diverse fronts. In general, land for farming is mostly in conflict with land demands for residential (housing) purposes in South Africa. This is also related to the major influx of migrants from rural areas. One member explained her perception of the overall problem and dynamics in the following way,

> the fact is the more houses you build the longer people become permanently urbanised and the rural areas are empty. Building houses is not going to solve the problem. You need to build farms, smallholder farms, small plots where people who are here already can uplift themselves.[39]

This statement relates to the fact that the provision of government subsidised housing (commonly known as RDP housing) is often outside the city centre, far from potential employment opportunities. Linking these households to agriculture is perceived as one solution.

Beyond that, KEF requests access to public land for food production. KEF has been interested in creating community gardens at the Old Crocodile Farm at the city outskirts between Thembalethu and Pacaltsdorp as well as in the buffer zone in Blanco. Moreover, the initiative has demanded official access to farmland in Thembalethu. These ideas and demands can be framed as conceived space (mental concepts). In this section, these three cases are outlined. While the first two cases are characterised by rejection from the official side, the case of Thembalethu tells a long story of neglect.

The municipality rejected the requests for community gardens at the Old Crocodile Farm and in the buffer zone in Blanco in 2015. The Old Crocodile Farm is located on the edge of the city centre and on the main route to Pacaltsdorp and Thembalethu. One of the founders was convinced about its suitability as a permanent site for KEF:

> It would be a perfect training ground for Kos en Fynbos. A permanent site for Kos en Fynbos. There is water there. [...] It is close. You can train people how to grow their own thing. [...]. The community would use that land wisely. [...] We do not need land to divide the people.[40]

When discussing possible land management on such a potential public site, it is clear that the KEF members would like to see clear responsibilities, more established organisational structure as part of their group, to avoid mismanagement and trouble which are considered as companions of food and land commons. While the old farm has been closed since 2007, plans to sell it to the highest private bidder or to agree on a profitable long-term lease have remained unsuccessful. The municipality moreover indicated that the land is needed for housing. The interviewee who was involved in the interaction with the municipality lamented, "Maybe they could put 20 houses. You could have provided houses to 20 people or 20 families. But you could have trained 10,000 people over ten years".[41] However, the municipality does not consider a public community garden run by KEF as an option. The municipality's decision has to be framed in the context of neoliberal urban development and hence market-driven thinking. The interest in real-estate investments, for instance for hotels, is part of the region's tourist focus and the sector's potential for growth (George Municipality 2016, 78; South African Cities Network 2014, 60); they are not interested in starting a community-led development project with uncertain outcomes and low returns.

Another dispute occurred about the use of the buffer zone in Blanco. Although apartheid ended in 1991, like in many other South African towns the buffer zone in Blanco still fallow land. According to KEF, a community garden there would have offered an ideal opportunity to bridge the divide between communities and races and create a vital space for interaction. Although KEF got in contact with different government bodies, they were not able to get access to the land. A member involved in these efforts described the situation and the government's management of the zone as follows:

> They chopped the land up in lots of small different zones and gave one to each different government national department. [...] So, that in the end, it is so difficult to consolidate the land as a complete zone and it remains unused. [...] We realised we are not going to anywhere with the buffer zone.[42]

The division of the buffer zone across different national departments causes a permanent land blockage. For both the buffer zone in Blanco and the Old Crocodile Farm the LED tried to facilitate land access but was not successful. Hence, it asked KEF to abstain from land occupations.

Land access for farming is also a problem in Thembalethu, particularly for farmers at the fringes of the township in the Sandkraal area. The land of the former Sandkraal farm (Erf 197[43]), with about 700 hectares, presents an ideal area for farming located between the fringes of the township and the coast (see picture below). The land is owned by the Department of Human Settlements, Western Cape Government. In the late 1990s, after the end of apartheid, the government expropriated the land from white farmers. The expropriation was part of the Proactive Land Acquisition Strategy (PLAS). PLAS focused on buying farmland and using it for housing and farming to support the poor. However, these policies only benefited a certain group of experienced and financially better-of farmers with time-limited lease contracts (Department of Land Affairs 2006, 18). Those who did not benefit were basically forced into squatter farming. These land occupations of state-owned land have been a sad reality for more than 20 years. In the process of land reform, it would have been possible to reclassify this land as municipal commonage and thus to provide land access to poor farmers. The national Department of Land Affairs introduced the Municipal Commonage Programme in 1997. However, George refused to make commonage available to poor residents, choosing instead to lease it for profit (Anderson and Pienaar 2003, 19). In the long run, this is in line with South Africa's neoliberal ideology fostering growth and modernisation in the agricultural sector. From this perspective, small-scale urban food production, so-called mini-agri-villagers are not utilising the land efficiently and therefore block capital accumulation.

In 2016, more than ten years after the introduction of the first land reform attempts, an expert of the Farmers Support and Development Unit mentioned that many farmers were struggling. Much of the equipment and infrastructure provided was stolen, not maintained well or vandalised. The interviewee resumed: "our department in the 2000s invested millions into that project as well and it all failed".[44] According to PLAS, those farmers who successfully farmed the parcels during the trial-lease period could acquire full ownership (Department of Land Affairs 2006, 10–11). However, many farmers failed to fulfil these requirements, have continued to stay and hope for further support.

According to the farmers, further training, input in terms of seeds, fertilisers, and capital, as well as access to markets, were still needed in 2016. The informality in land access is closely tied to further challenges like inadequate access to infrastructure particularly water, proper roads, and storage facilities. One of the KEF leaders summarised the situation as follows:

> People who want to farm in Thembalethu face different issues than those in town. In town they have water, land, security, fencing, everything. Here we have pressure with urbanisation, more and more people are coming building shacks, there is less space for farming. We need official access to the land.[45]

The difficult conditions are further complicated by an ever-increasing influx of new arrivals; the number of squatters has been increasing, alongside those leasing the land. Thus, both groups – those who have been leasing (under PLAS) and those who have been illegally occupying land – were struggling to make a living. Both groups have been at risk of eviction. In response, farmers organised protest actions. In 2012 and 2013, particularly, these protests gained wider attention as they served as a reminder of the dispossession under the 2013 Natives Land Act and its dire consequences. The *George Herald* described the following: "About 50 small-scale farmers participated in a protest march [...]. They are protesting against their eviction from land in Thembalethu", and a local farmer emphasised "we plant vegetables for own consumption, but we cannot develop our farms further because of the ongoing dispute" (2012). In the past, the government offered those who leased the land purchase agreements but abstained from materialising this offer later. These overt protests can be seen as an example of O'Brien's notion of rightful resistance, as the government withdrew from official agreements and thus showed disloyal behaviour (see O'Brien 1996, 33).

The precarious situation of the squatter farmers gained fresh attention in 2014. New hopes were sparked amongst the squatter farmers when the Western Cape Department of Human Settlements introduced the Sandkraal Integrated Framework for Development targeting the needs of the squatter farmers and the community as a whole. Two years later, the framework became an integral part of the newly introduced Thembalethu Precinct Plan for Urban Upgrade (CNdV Africa 2016). However, there has been no evidence of attempts to improve the situation of the farmers in recent years, and most of them still do not have official land access.

Sipho, one of the KEF leaders, represents the Thembalethu Farmers Unity (Umanyano), which mainly cultivates fields in the Sandkraal area. Sipho explained that they have been struggling to get official land access and improved access to water since 2000. The unity introduced its request and proposal for land use to the municipality and Western Cape Government several times, with the aim of being incorporated into the government's development plan for the area. He mentioned: "This land access thing comes to the point of politics and business this is my feeling. [...] Things should go more socially. People should talk and raise their voice in the debate".[46] The *George Herald* regularly reported on the situation of the farmers, tried to follow up with the government and did short interviews with representatives. A journalist summarised the farmers' land demands:

> they [farmers] want that space to be formalised – the usage of the certain sections of that land – it's a huge part of land. So that they can plant their veggies and keep their livestock [...] But government must decide what they are going to do; they can't just squat there and put up their shacks.[47]

Again, this shows the notion of O'Brien's rightful resistance and the support by "influential advocates" (O'Brien 1996, 33).

The perceived silence of the government permeates people's everyday lives and fosters further exclusion; but it also strengthens community ties. However, this negligence and precariousness are not accepted silently; people continue to occupy farmland and engage in street protests. Sipho considers KEF as a driving force and platform to address these issues in public beyond the traditional boundaries of the community.[48] This cooperation between the different KEF members in the communities clearly relates to food sovereignty's aim for "positive interaction between food providers in different regions" (Nyéléni 2007). While members jointly demand official access, they are rather interested in individual land usage as proposed by the government in several development plans. Clearly, a commons perspective bears its own challenges and potential conflict within communities.

One farmer reflected critically on the machinations of the authorities involved (Human Settlements, Agriculture, Public Works, Transport, and Water and Sanitation) regarding land access:

> the land is getting smaller and smaller. What is it? Urbanisation and privatisation. Every piece of land we identify there is a dumping area there that of the people, and we fought with municipality. [...] Me, I am not playing with the people, and I don't want someone to play with me too. [...] So where is this going?[49]

While the ultimate demand of these farmers is official and long-term land access, this also connects to further demands and silent struggles on different fronts: access to farming equipment, extension services, infrastructure, affordable and healthy food, and labour market integration. These unfulfilled and neglected demands shape the conceived space, i.e. the idea and mental constructions of space, as limited and exclusionary with a lack of official titles. Such developments are in line with O'Brien's overt and rightful resistance. He argues that, as long as there is a gap between what is promised (end of apartheid and land redistribution under PLAS as well the Integrated Development Plan) and what is delivered, mobilisations invoking people's rights will become the node around which people mobilise (O'Brien 1996, 45).

While the material space, the land and water, people's crops and harvest, are concrete and "real", the mental part and the conceptualisation of space, are restricted by the fact that they might have to leave the land and would be excluded from all the resources that they need to survive. It is a permanent state of insecurity. At the same time, out of these critical examinations, people carve out emergency solutions which have, over the years, turned into a sort of permanent solution. All this is also reflected in the spatial practices, i.e. the material space, the restricted land and water use. The third element of space, the spaces of representation, the lived realities on these occupied lands, shows that these land users have created and strengthened an identity which is closely connected to the land and which has empowered them to fight for it and organise themselves. The land occupations of the farmers might be considered a way of enforcing food sovereignty's "right of local communities to inhabit and use their territories" (Nyéléni 2007, 1). These dynamics of creating space and alternatives to overcome exclusion refer to autogestion (Lefebvre 2002, 779–780; Purcell 2014, 150). People manage space for themselves, ignore the restrictions put in place by official bodies, and in this way actively build and shape their lived space. This refers not to legal ownership and the commodity value of the land, but rather to the continuous use value of the land which facilitates access (Purcell 2014, 142).

Besides the land access demands of the farmers' group, land pressure is also caused by institutional land needs, e.g. for schools by the Department of Education, in Thembalethu. This was explained by a community worker:

> I got one commercial [farm] there but now education need that place for the school. […] Now I already talked to these people [the farmers], if education says wake up, now I am coming, you must pick up your things and go. […] It was 2004, but now that ground is for education, it is not for to be plant.[50]

As a result, this group of farmers had to leave the land and intended to join the land occupations of the other farmers in the Sandkraal area.

The responses of several government departments have failed to meet the needs of the smallholders. Government attempts to push economic growth through tourism and small-scale farming operations are difficult to operationalise in a full-blown township with its security issues and a vital informal economy. It is particularly concerning that governmental interventions are maintaining uncertainty over a long period of time. The proposed Integrated Framework for Development has always been quite unlikely to be implemented for various reasons. The local departments involved seemed over-burdened and lacked the necessary financial means to bring the plan to fruition. Furthermore, the responsibilities of the different departments (i.e. Human Settlements Department, Department of Agriculture, George Municipality, and Rural Development Department) are unclear. On the whole, municipalities are overwhelmed by the situation in townships and particularly the transitional nature of informal settlements.

The three cases examined above demonstrate a range of difficulties in making land available for food production. The relevant government departments and the municipality seem unable to balance different land demands (i.e. for housing, farming, infrastructure, commercial purposes). However, these lived realities remind us of the importance of engaging seriously with the urban land question in the ongoing land reform debate in South Africa. The experience of land occupations in Thembalethu leads us to question whether equal land access can ever be the solution. Based on my empirical findings the answer is no. While land access was provided for some farmers by PLAS, it did not resolve the problem of marginalisation, which is the main issue for all land squatters in the area. It is thus essential to figure out how livelihoods can be made secure in a context of marginalisation.

In South Africa, struggles for land access tend to have disappointing outcomes. This is true of the Old Crocodile Farm and the buffer zone in Blanco, where land access demands were rejected after a short period of time, and where KEF did not seek direct confrontation with the municipality and government departments. However, the case of land access in Thembalethu is different; here, people have been requesting land access for more than 20 years, and several promises have been made by the government. Still, no clear decisions have been forthcoming. While creating visibility for these issues, KEF prioritises the people and their needs and therefore explores what might be possible politically.

Community knowledge and skills

The fifth pillar highlights that "Food sovereignty builds on the skills and local knowledge of food providers and their local organisations that conserve, develop and manage localised food production [...] and passing on this wisdom to future generations" (Nyéléni 2007, 1). Valuing traditional and tacit knowledge plays an important role in this regard (Martínez-Torres and Rosset 2014). In this way, a knowledge base should be constructed jointly, be accessible for everyone and open to diverse layers of knowledge in favour of biocultural diversity. Another critical aspect of knowledge is so-called critical food literacy. According to Yamashita and Robinson this implies, "[to] grapple with multiple perspectives and values that underlie the food system, understand the socio-political contexts that shape the food system, and take action toward creating just, sustainable food systems" (2016, 269). They identified sensitive issues like the impacts of GMOs or climate change.

The appreciation of community knowledge is evident in KEF's work in different ways. Several members have indicated that there is a rich body of traditional and indigenous knowledge in their communities. Highlighting the second part of the group's name, fynbos, KEF members know a lot about herbs and medicinal plants. Beyond that, the discussions and interviews revealed a range of skills in planting, preservation of nature, preparation and conservation of food. All these are different kinds of local and community knowledge. Various ways of uncovering and sharing this knowledge in the group amount to an attempt to create awareness and knowledge conservation. At the same time, engagement with these diverse layers of knowledge also points to a repertoire of protest – a kind of struggle against the neglect and consequent disappearance of local, traditional, social, and environmental values, which often have deep roots in the communities. This section starts with a brief overview of the perceived decline in local knowledge. It continues with a sketch of the group's engagement with the construction of knowledge and the struggle against its decline.

According to the initiative, the conceived and perceived space of knowledge in food cultivation, preparation, and health is characterised by a damaged connection to nature and food. In this sense, food also implies identity; it builds from cultural as well as social and ecological values. Different stories show that knowledge assets tend to get lost. These seem less appreciated and needed in a modernised world, which increasingly relies on technology and "efficiency". For instance, members described that with the wider availability of commercial medicines, medicinal plants have become a niche area. At the same time, members indicated that knowledge of food production and preparation becomes less relevant as neither is deemed necessary for survival. Against this background, the education system pays little attention to cooking and gardening skills. NGOs or community organisations often seem the only actors providing inputs in these fields. Migration is another factor contributing to changing food choices and declining practice of certain skills, including farming. Changing lifestyles and different labour engagements can shift knowledge which was once indispensable into a hidden corner. Moreover, these knowledge shifts have partially historical reasons, as one member explained to me: for instance, the political dominance of white people in the past strongly influenced local peoples' perception of indigenous knowledge:

> In South Africa, we have a huge diversity of plants, [...] everything is used for medicinal things, and when the white people first came here, they actually learned from the indigenous people to use these things, and a lot of the white people just took it upon themselves and even call it bure rat.[51]

Bure rat is Afrikaans and can be translated as farmers' knowledge. This experience can be considered as colonisation of knowledge and relates to a hierarchy in knowledge creation and application. The initiative's intention is not to blame, but rather to (re) create a kind of pride and use value of local knowledge in general. The lived spaces of the different members expose and dismantle these knowledge shifts and losses. Besides addressing these issues, the initiative is also actively involved in expanding members' knowledge base, valuing diverse kinds of local knowledge, and proposing different ways of applying it in daily practices and in this way preserving it.

Daily practice of gardening and food preparation makes a rich contribution to mutual learning. An employee of the George Municipality acknowledged the everyday practice of sharing of which knowledge is an essential part: "They [KEF members] will share their knowledge [...]. For me, coming from where I am coming, this was an eye opener. It is bringing you back where you are coming from".[52] This person is from Blanco and in this context remembered old times, where connections between the neighbours were closer and more supportive.

The philosophy of permaculture plays a key role in this regard, shaping the KEF's farming practices and inspiring its attempts to share knowledge and to use natural resources in a sustainable way. One member explained: "The whole permaculture concept is to care for the earth and care for the people and share, share, share. [...]. Everybody was grassroots level, and everybody helped everybody else".[53] For instance, in several meetings Spekboom seedlings (*Portulacaria afra*) were shared as this tree helps to counteract carbon emissions. One of the initiators, who had moved to Blanco, admitted, "I was learning from them [the other gardeners]. I didn't know the plants; I didn't know how to make compost".[54] The coordinator in Thembalethu reflected on her exchange with other members:

> There were things that you were not even aware of them. You were not even aware which plant to plant in which season. So, as soon as you go to this group, they open your eyes, this is the time to plant it. This goes well with that. So, I think with this kind of information they are going places.[55]

For KEF, sharing knowledge on an equal footing is taken for granted. One member demonstrated this simple daily practice in her community: "Okay well, you have a wattle fence and you know how to do that, I know how to prune fruit trees. So, let's get together".[56] To make knowledge about food cultivation widely available, KEF member developed a double page for the *George Herald* with a calendar and tables of seasonal vegetables and fruit trees and plants that are recommended for any garden. These were published in September and October 2015 and also provided as printouts. In addition, illustrations showing how to prepare compost were provided.

Beyond these practices of sharing, the continuous extension of their knowledge base takes place on four overlapping fronts: (1) legitimisation of everyday knowledge, (2) revitalisation of past knowledge, (3) preparation of the future generation, and (4) integration of external expertise. Interventions on these fronts contribute to the construction of identity and self-sufficiency. They are outlined in the following paragraphs.

(1) Legitimisation of everyday knowledge Grounded in everyday life, the initiative fosters appreciation of everyday local knowledge particularly regarding herbs and partly in food preserving, which often remains forgotten or hidden. A member explained that quite often, "people say, I don't know anything about herbs. [...] you ask them what is this plant? And then, they know. And then, they say, yes but it is not part of my everyday practice. This is also what we are doing with KEF. [...] It is kind of taking this grassroots knowledge and making it important again".[57] This refers for instance to medicinal usage of certain plants in case of earache or stomachache and avoiding unnecessary visits of a clinic.

Besides empowering people to use these skills and knowledge, the group promotes its interest in self-sufficiency at least for basic medical treatment. In fact, many residents, especially the older ones, already use herbs. The coordinator in Blanco proudly claimed, "when we grew up and we never go to the doctor".[58] Some members said that they see their garden as a pharmacy. For instance, the perdepis tree (*Clausena anisata*), a medicinal plant, seems to be a standard in KEF gardens; it helps with different ailments. It is moreover typical for many KEF members to brew tea with home-grown herbs (e.g. chamomile, lemon verbena, sweet root). This is also the case for food preparation. Different experiences and tricks are shared and promoted in the group, and thus have become part of its knowledge base. What is more, medical and food preparation knowledge passed from previous generations – some of which can be traced back to indigenous tribes – has a specific value. This is closely linked to the revitalisation of past knowledge.

(2) Revitalisation of past knowledge As well as medicinal knowledge, skills in animal rearing or honey making are passed from generation to generation. While those skills are not necessarily "hidden", this knowledge relies on experience, learning by doing, practical application – so-called tacit knowledge – which might be considered as informal. For instance, the collection of wild fynbos honey in the forests can be seen as tacit knowledge, which has deep historical roots in the area. People do not acquire these skills in school. Moreover, knowledge on food processing but particularly gardening was often passed directly from the parents or grandparents. In this regard, it is another intention of the initiative "to train people how to do bottling and preserving. That's a skill that the old people have. But the younger people like the office workers and teachers don't have that skill. They didn't learn from the grannies. [...] We lost all those skills, and we need to relearn them. How to look after ourselves".[59]

In the three communities, many indicated that their families had roots in farming and gardening and passed on their skills. This was also indicated in several newspaper articles, for instance "My parents were strictly spoken peasants" (George Herald 2014b, translated from Afrikaans). Many of the food producers in Thembalethu left the countryside (mostly the Eastern Cape) because of limited job opportunities. People in Pacaltsdorp and Blanco either have been living there for a long time or moved within the Western Cape region. In these areas, most of the inhabitants or their direct family members had connections to food cultivation or livestock raring. KEF is resuscitating this knowledge and these skills.

The elderly coordinator in Blanco indicated that she became inspired by a Khoisan who showed her and others medicinal plants in the forest. She shared the following experience: "they always leave some of it [herbs]. But in our days, if we take herbs, we take the whole bunch. Not that people. We can go and learn much about mother nature with that people. They were the first people the Khoisan because they know how to live and how to survive".[60]

One member highlighted the continued connection to the land of many South Africans and its continued importance for identity, despite forcible removals and broader experience of exclusion from the land in the past:

> their real connection to the land has been damaged in many ways, I do believe that we can get that connection back. [...] the grandfathers or their grandmothers did actually grow food. I think it is very empowering to find out [...] more about your identity [...] and then to do it again to help yourself.[61]

The KEF coordinator in Thembalethu is convinced about the knowledge inherent in this connection to the land and its usefulness for the gardeners and farmers in her community: "Many do have the knowledge. But they don't have a place to apply the knowledge".[62] While KEF's work reveals that some people are ready to use the knowledge passed from previous generations and stick to their identity, the case of Thembalethu reminds us again that many township dwellers do not have sufficient resources; they particularly lack official land access to apply their skills and in this way sustain their livelihoods.

(3) Preparation of the future generation One of the intentions of food sovereignty is passing "wisdom to future generations" (Nyéléni 2007, 6). KEF members try to put that into practice. Although KEF has been actively reaching out to people, the group also realised that the younger generations are difficult to target and rarely engage in gardening. It is their intention to empower young people who are unemployed to engage in food production for several reasons. Hence, the younger KEF coordinator in Blanco explained: "we got to teach our youth again how to be independent, because they don't know gardening. Just know technology. That is all their education how to survive. We have to get the youth as actor".[63] The overall intention is to provide practical examples and knowledge of gardening to the next generation. The coordinator in Thembalethu underlines these intentions in her community. She describes how gardening skills might play an important role in the context of high youth unemployment:

I have three daughters the middle one is always in the garden with me. […] I also involve them, because at the end of the day you don't know where you gonna end up in life. […] even if there is one day there is no job at least you can grow your own food.[64]

Part of this ambition has been to initiate gardening workshops including the set-up of vegetable and herb beds in the university, schools, and crèches all over town. With several activities the KEF reaches out to younger people and tries to get them on board. This includes involvement in local events like the national science week and garden tours for students in the communities involved.

KEF members also mentioned that many young people do not know how to prepare meals from fresh ingredients. Thus, first attempts were made to teach the preparation of simple, delicious, and healthy meals from the harvested produce. In this regard, one of the members shared her overall concerns about the dependence on technology and indicated the need to learn at least the basic survival strategies:

What people are going to do when the drought is going to get bad? What is if the load shedding keeps going on? We didn't know how to make a fire and if there are no matches. […] That are all the things that we must bring back and tell the people because our forefathers didn't need that things.[65]

In this context, it has to be acknowledged that the Western Cape region has been affected by several severe droughts and phases of load shedding during the last four years.

(4) Integration of external expertise While a huge part of this knowledge originates from the communities themselves, specific skills on permaculture, like compost-ing, lasagne gardens, or companion planting were particularly introduced and strengthened by those working with NGOs such as the Landmark Foundation or Permaculture South Africa, or in the field of education such as NMMU. On an informal and mostly voluntary base, KEF is part of these outreach activities.

Recent interventions of the NMMU (Soil Science Department and Sustainability Research Unit) are described in this way: "they were quite happy to provide skills and knowledge transfer. So, they ran a workshop last time – a knowledge sharing workshop – very well organised. They did a compost workshop [...] mainly tackling the skills and knowledge".[66] A lecturer of the Sustainability Research Unit, NMMU, and her students have also recognised these locally created knowledge assets. They decided to learn from KEF and to make this knowledge widely available. Hence, they invited about 20 members for a workshop at the university in 2016. It was mostly elderly people who joined the session. The gardeners shared and discussed. All these outputs were summarised in the so-called Kos en Fynbos Encyclopaedia, which was made publicly available in the same year. In a discussion with the responsible lecturer, she particularly referred to the rich tacit and traditional knowledge evident in the group.

In addition, several members promoted a critical reflection of the commercialised agriculture system. This included, for instance, critical debates about GMO seeds, long travel distances of food, and the increasing penetration of supermarkets. This was partly done by those who are better educated and those who are working in the field of education. One of the initiators and one leader have moreover been active in sharing this kind of knowledge broadly on KEF's Facebook page. They have shared newspaper articles, informative website posts in the broad field of sustainable agriculture, and DIY gardening practices. Related posts are intermixed with short stories and photos of KEF's gardening demonstrations and interventions. All this is a form of advertisement reaching out to the broader public and it fuels the critical knowledge base of the initiative. In this regard, O'Brien observed that "increased mobility and media penetration have made them [resisters] more knowledgeable about their exploitation and about resistance routines devised elsewhere" (1996, 41). It is clearly visible that social networks and the media also help social movements to formulate concerns and demands.

In sum, the outlined lived experiences point to the continuous expansion of different layers of knowledge and skills with their inherent social and ecological relations. As illuminated with the critical urban food perspective, most of the information and expertise is not necessarily taught in the education system and thus exceeds common knowledge. The group places a high value on mutually created knowledge.

Biodiversity and connection to nature

The sixth pillar of food sovereignty focuses on the interconnection with nature. In this sense, food sovereignty builds from "the contributions of nature in diverse, low external input agroecological production and harvesting methods that maximise the contribution of ecosystems and improve resilience and adaptation, especially in the face of climate change" (Nyéléni 2007, 1). Consequently, industrial, high-input agriculture, which is harmful to the environment is considered critically. This pillar is strongly connected to pillar 5, which focuses on local knowledge. Agroecological production is a key element of food sovereignty (Martínez-Torres and Rosset 2014). This refers to an ecological rationale, small-scale production, biodiversity, and traditional knowledge (Altieri, Funes-Monzote, and Petersen 2012, 2). In this respect, the concept of permaculture is gaining popularity (Ferguson and Lovell 2014, 251). Again, many parallels are evident between these ideals and KEF's work. In a meeting in 2016, some members pointed out that they would call themselves environmental activists. The activism and lived space are related to the perceived and conceived destruction of the environment and related negative consequences. This section thus briefly frames related problems described by KEF members. The second part highlights their interventions, which illuminate possible alternatives.

In general, KEF has been critically observing the increase in large-scale agriculture, weather extremes, and the loss in biodiversity. Some members are well informed about the environmental degradation in the region, e.g. decline of animal species as well as environmental and industrial pollution. A group of engaged gardeners sketched out a decline in the variety of fruits, vegetables, and herbs grown and used in their communities. While the initiative truly appreciates the nature in the area, "It's Eden, it's fertile, it's green"[67], there is a lot of environmental sensitivity visible in the initiative's discussion and daily practices: "All of this [KEF's work] is to protect mother earth".[68] A gardener from Thembalethu summarised, "When we think of our mother nature, I think we the Kos en Fynbos folks agree, a lot of damaging is taking place in our communities but also out there".[69] Overall, many members emphasised the high value and appreciation of nature through the terms "mother nature" or "mother earth".

Members moreover linked the perceived environmental degradation to an increase in different kinds of sicknesses, which is all part of the motivation of growing food in an ecologically sustainable way. On the one hand, a large variety of home-grown fruits, vegetables, and herbs are perceived as essential in the process of healing and staying healthy. On the other hand, the industrialised food system is considered one of the issues behind ill health. This is exemplified in a personal experience of one member, who grew up on a farm north of South Africa and one relative was diagnosed with breast cancer. Another member shared: "One of the research people in this area said, why has this town such a high rate of cancer and did research, and then they found it were all these farm children who were running behind the crop sprayers. This is how [she] in fact first knew about the dangers of these insecticides and pesticides and the genetically modified things and Round-up".[70] Consequently, such lived experiences have been encouraging people to make a difference, to abstain from chemical fertilisers and grow their food naturally, which the group also coins as traditional farming. These experiences also fuelled their concerns about large-scale agriculture. Another member reflected, "large-scale agricultural farming that are actually the ones which are causing the destruction of our biodiversity".[71] A farmer in Thembalethu further detailed the negative health effects of chemical fertilisers and refers to the experience of some farmers in the Sandkraal area:

> We need biosafety not chem, it is dangerous. [...] For some they took those chemicals if they got it from a cooperation by this agricultural fund and this is for their system. And those people are not trained to use them [...]. Then they can get wounds [...]. So, now we encourage the traditional system.[72]

He was referring to government interventions and extension services through which these inputs were supplied in the past. However, detailed instructions, follow-up, and monitoring were not provided. For this farmer, a traditional system implies farming without chemical inputs. In this way, the initiative calls for working hand in hand with nature to protect natural resources and health. This is also reflected in the practices of some members who were inspired by the Khoisan, who focus on protecting nature and only harvesting as much as needed, and thanking nature for the goods it provides. These practices and intentions are in stark contrast to the country's focus on neoliberal development and technical solutions to feed the nation (see Chapter 3). According to the group, many South Africans – including the ruling elite in the field of agriculture – are not in favour of protecting the environment. One of the coordinators in Blanco thought out loud, "All the rich people who live like high up, they don't care for mother nature. […] There is nothing they can buy high up there. They must come back to earth. We must all come back to earth even our president. […] You can't live if you haven't got veggies, if haven't got water".[73] The overall critical stance regarding highly industrialised agriculture led members to avoid chemical fertiliser and GMOs in their gardens and in food consumption. Related knowledge and negative experiences are widely shared in the group and cause subtle acts of everyday resistance.

The group aims to rebuild the connection to nature. Hence, the concrete and material space of the garden or farm plot can be framed as a so-called space of hope in which ecological sustainability is of importance. Several discussions and newspaper articles refer to KEF's focus on biodiversity, particularly valuing and protecting the ecosystem and its variety. Moreover, several interlinked and overlapping daily practices fit under the broad frame of permaculture farming for biodiversity, for instance, recycling and integrated waste management including composting; companion planting; no-dig gardening. Members vary slightly in the terms they use to describe their cultivation practices. While the term permaculture is mostly used by those working in the field of education, with NGOs or those running the gardening demonstrations, the applied practices still fit under a holistic frame of sustainable agriculture. Daily practices present many parallels to the principles of agroecology, which is an integral part of food sovereignty (Altieri, Funes-Monzote, and Petersen 2012). However, the initiative has not been using the term agroecology.

The broad frame of permaculture and biodiversity refers to the lived practice of growing different plants that go well together, diversifying and rotating crops, and avoiding external inputs (i.e. synthetic fertilisers and pests). Such conditions are to be found in most of the gardens and thus might be considered as the minimum criteria. Some people diversify more than others. For instance, farmers in Thembalethu tend to vary only two or three crops, while some of the backyard gardens present a kind of permaculture forest. Mixed planting is also part of companion planting, which groups crop species together that, in combination, have beneficial natural characteristics for instance in pest control, crop productivity, or attracting insects. Furthermore, integrated planning is evident in many gardens. Diverse resources are used, for instance water from the neighbouring stream, or harvested rain water, while some even grow a vegetables on the wall or roof of their house (e.g. melons, tomatoes, pumpkins). Basically, permaculture implies avoiding monoculture cropping, and diversification of the garden. In this way the soil regenerates quickly, which allows the cultivators to eat from the garden all year round and not only in one season. A passionate gardener explained:

> you grow just enough that you can eat maybe just a head a week. […] It is like eating seasonally, eating diversely […]. And then also to learn what puts goodness into the soil. Like people look in that garden and say but you can't eat that herb over there, but the herb attracts bees.[74]

External synthetic inputs are replaced by environmentally friendly sprays against pests. Many also produce organic fertiliser. A farmer from Thembalethu explained:

> There are some people who know how to make natural fertiliser with traditional tobacco. […] there was a traditional and healthy agriculture […] it all disappeared with the chemical stuff. […] Those who distribute that chemical stuff do it for their own benefit not to rescue the people from poverty.[75]

He indicated that many of those who are farming in Thembalethu are still using these traditional practices, not least because it is less costly.

Composting is used to recycle organic waste in the garden and to improve soil. Making a proper compost heap is one of the KEF demonstration elements. Using manure from farmers in the neighbouring villages is another integral element. Some regularly collect manure and bring it to town, particularly to the resource sites.

Recycling of inorganic material is rather a niche activity, but some people use plastic bottles as borders for vegetable beds or as flowerpots. They use water from the washing machine and from the kitchen to water plants. Other know how to build stoves out of tin barrels. With an increase in illegal dumping in George, KEF has also been interested in stepping up its efforts in waste management and supporting the municipality in this, which also forms part of the fenced-in resource sites.

Sustainable seed supply is another element of KEF's work. One member explained: "we are going to teach the people seed saving. Because part of permaculture is that you are permanently in supply of your own seeds".[76] However, seed saving efforts are still at an early stage; the majority of the members still rely on external seed donations from other members or KEF donors. It also quite common to share seedlings.

There are certainly members who are more used to organic farming than others, some of them are keen to apply the permaculture principles more comprehensively whereas others still need more training. One of the very passionate members was convinced that, "You can teach people how to do natural building. How to build with straw and clay. Like the house I lived in in Blanco. You can train people how to do that".[77]

In sum, KEF breathes life into an organic and diverse way of food production to protect the environment, encourage healthier eating, to benefit from nature and avoid destruction of the environment. This form of food production is also suitable, given the restricted financial and technical means available to grow food in these communities.

Concluding remarks

Overall, the KEF initiative offers a breeding ground for various transformative efforts and self-help strategies. Using the critical urban food perspective, this chapter shows the initiative's attempts in fighting broader inequalities in the city and in the agri-food system and, in this way, creating alternative food politics mirroring food sovereignty. The group facilitates visibility through materially spatial aspects, such as growing food and sharing seedlings and produce. In addition, it engages in discursive elements, such as knowledge conservation and exchange as well as critical reflection of development projects. More specifically, this chapter explored the food sovereignty pillars on the ground and added local particularities. The final part provides a short summary of the related five sections and highlights key take-away messages.

First, efforts towards local production and consumption are clearly visible in the initiative's work, which mirror pillar 1 and pillar 3 of the Declaration of Nyéléni. Members reported that they support the local economy and prefer to buy in small house shops and food stalls. However, hectic working days as well as an increasing availability and acceptance of processed food contributed to unhealthy diets. The experiences of several participants are key to understand the rapid nutrition transition and its impacts in deprived areas. Given a critical stance towards the prevailing nutrition and farming conditions in the host communities, KEF actions in strengthening the local food system are diverse, reaching from the cultivation of traditional crops to diverse practices of produce and seed sharing. In general, the initiative unites concerns and needs of consumers and producers; both sides are enshrined in the lived practices of the group.

Second, appreciation and empowerment of local food providers is another key feature which relates with food sovereignty pillar 2. It is particularly the government's work in facilitating food production activities in the region, which is perceived critically by KEF. There is a janus-faced aspect to governmental interventions. On one side, the municipality and government departments appreciate the ongoing gardening activities and incorporate them into their own projects. On the other side, there is evidence of governmental neglect of the initiative. In general, the local government is over-burdened with the fallout of high unemployment rates, fails to adequately support self-help strategies, and to introduce public policies facilitating healthy food provision beyond the power of the market. The group basically exposes these failures on the part of the government and shows that its members' reliance on farming is also not fully recognised by the authorities, which fail to provide land access and adequate support, for instance in terms of market access to food producers. In response to this lacking support, several members occupy land, strengthen informal food networks and market structures. These informal structures and networks parallel with Scott's everyday forms of resistance and Bayat's notion of quiet rebels – the engagement in structures which are not necessarily beneficial to the formal economy and in several cases remain invisible for outsiders, apparently even the government. Thus, ties to the informal economy might even be interpreted as a reminder of the need for more appreciation or integration of economies from below into the formalised economy. Informal structures point to social capital and the creation of alternatives to the restricted and exclusionary formal economy as well as to the lack of social protection. Only by considering and valuing these structures can the government advance any form of meaningful support and engagement with them.

Third, KEF has been contributing to the localisation of control over resources required for food production, e.g. seeds, knowledge, and tools, tying in with the forth pillar of food sovereignty. However, land access remains one of the essential demands of the group. Although many members grow food in their backyards and in some areas on public land, secure access to land remains challenging, particularly in densely populated neighbourhoods. In this way, the material space is limited. While several local government bodies have made different interventions in fighting land inequality in the aftermath of the apartheid era, urban land demands remain neglected. Being affected by these inequalities, KEF has been active in exposing the lack of consideration for land demands on diverse fronts. In general, land for farming is mostly in conflict with land demands for residential purposes or real-estate investments. In this context, KEF's experiences clearly show the restrictions maintained by the government and limited bargaining space of the group, which apparently can only be targeted with rather radical interventions of land occupations. This is illuminated by several requests for access to public land for food production by the initiative. KEF has been interested in creating community gardens at the Old Crocodile Farm at the city outskirts between Thembalethu and Pacaltsdorp as well as in the buffer zone in Blanco. Moreover, the initiative has demanded official access to farmland in Thembalethu. While the first two cases are characterised by rejection from the official side, the case of Thembalethu tells a long story of neglect.

Fourth, the initiative is valuing community knowledge and skills in food provision. Notions of the fifth pillar of food sovereignty are mirrored here. Several members have indicated that there is a rich body of traditional and indigenous knowledge in their communities. Highlighting the second part of the group's name, fynbos, KEF members are well informed about herbs and medicinal plants. Beyond that, the discussions and interviews revealed a range of skills in planting, preservation of nature, preparation, and conservation of food. All these are different kinds of local and community skill and expertise. Various ways of uncovering and sharing this knowledge in the group amount to an attempt to create awareness and knowledge conservation. At the same time, engagement with these diverse layers of knowledge also points to a repertoire of protest – a kind of struggle against the neglect and consequent disappearance of local, traditional, social, and environmental values, which often have deep roots in the communities. The construction of knowledge and the struggle against its decline are considered as key elements of food sovereignty construction. According to KEF, the conceived and perceived space of knowledge in food cultivation, preparation, and health is characterised by a damaged connection to nature and food. In this sense, food also implies identity; it builds from cultural as well as social and ecological values. Different stories show that knowledge assets tend to get lost. These seem less appreciated and needed in a modernised world, which increasingly relies on technology and efficiency.

Fifth, the group aims to rebuild the connection to nature and thus several ecological components are visible in their work for food provision. Related aspects are clustered under pillar 6 of food sovereignty. On the whole, the concrete and material space of the garden or farm plot can be framed as a so-called space of hope in which ecological sustainability is of importance. Several discussions and newspaper articles refer to KEF's focus on biodiversity, particularly valuing and protecting the ecosystem and its variety. This is made visible by several interlinked and overlapping practices, which fit under the broad frame of permaculture farming for biodiversity, for instance, recycling and integrated waste management including composting; companion planting; no-dig gardening. Daily practices present many parallels to the principles of agroecology, which is an integral part of food sovereignty.

While the initiative comprises diverse imaginations of sharing, living in harmony with nature and neighbours, specific preference for land and food commons is not visible and even a bit of cautiousness is evident. This can be linked to broader doubts about possible mismanagement. Certainly, a commons perspective bears its own challenges and potential conflict within communities. Similarly, several food sovereignty proponents and activists do not call fiercely for a re-imagination of food or land from a commons perspective (cf. Holt-Giménez and van Lammeren 2020). It is essential to remember food sovereignty's roots in this context. Basically, calls for food sovereignty are strongly tied to demands of deprived (e.g. landless) small-scale farmers who are struggling to access markets and compete with many big players along the food value chain. One can argue, that marginalised small-scale producers are not necessarily interested in food or land commons. Individual access rights still play an important role.

Overall, the interventions of KEF present many entry points and clear linkages to food sovereignty in discourse and practice. It is important to keep in mind that food sovereignty has its origins in locally initiated responses of farming peoples, whose multi-faceted experiences breathe life into the construction of food sovereignty in different settings. Using different notions of space and resistance, as outlined in the introduction, proofed to be essential in shedding light on these transformative practices. Although the focus is on a single case study in South Africa, these realities and observations are of interest to understand diverse socio-economic and agrarian developments at the peri-urban fringes in other parts of the world. The specific findings allow to develop the existing food sovereignty pillars further, to carve out claims, and to adjust these in the specific urban-rural setting. This is done in the following chapter.

Notes

1 Interview on January 15, 2016.
2 The focus of the survey was purchasing behaviour; it was conducted in five communities in George (Blanco, Pacaltsdorp, Uniondale, Thembalethu, and Touwsranten). A total of 7,148 people participated.
3 Interview on March 10, 2016.
4 Interview A on August 23, 2016.
5 Interview on March 13, 2016.
6 Interview on January 15, 2016.
7 Interview A on August 23, 2016.

8 Interview on August 24, 2016.
9 Interview B on March 11, 2016.
10 Interview A on August 23, 2016a.
11 Interview on August 16, 2016.
12 Material spaces also comprises those elements which are determined by capitalism, for instance available groceries offered by dominant retail chains.
13 Interview on March 6, 2016.
14 Interview on January 15, 2016.
15 Interview on March 6, 2016.
16 Interview on March 6, 2016.
17 Interview A on March 11, 2016.
18 Interview on August 20, 2016.
19 Interview on August 19, 2016.
20 Interview on August 19, 2016.
21 Interview on August 19, 2016.
22 Interview on August 19, 2016.
23 Interview on August 16, 2016.
24 Interview on August 20, 2016.
25 Interview A on March 11, 2016.
26 Interview C on March 9, 2016.
27 Interview on January 15, 2016.
28 Interview C on March 9, 2016.
29 Interview on January 15, 2016.
30 Interview on January 15, 2016.
31 Interview on January 15, 2016.
32 Interview B on March 11, 2016.
33 Interview on January 15, 2016.
34 Interview on August 16, 2016.
35 Interview B on March 10, 2016.
36 Interview on August 16, 2016.
37 Interview A on August 23, 2016.
38 Interview on March 6, 2016.
39 Interview on March 6, 2016.
40 Interview on March 6, 2016.
41 Interview on March 6, 2016.
42 Interview on January 15, 2016.
43 Erf is Afrikaans for plot.
44 Interview on August 19, 2016.
45 Interview on August 20, 2016.
46 Interview on August 20, 2016.
47 Interview A on August 17, 2016.
48 Interview on August 20, 2016.
49 Interview B on August 23, 2016.
50 Interview A on August 23, 2016.
51 Interview on August 16, 2016.
52 Interview C on March 9, 2016.
53 Interview on March 6, 2016.
54 Interview on March 6, 2016.
55 Interview A on August 25, 2016.
56 Interview on January 15, 2016.
57 Interview on August 16, 2016.
58 Interview A on March 11, 2016.
59 Interview on March 6, 2016.
60 Interview A on March 11, 2016.
61 Interview on August 16, 2016.

62 Interview A on August 25, 2016.
63 Interview B on March 11, 2016b.
64 Interview A on August 25, 2016.
65 Interview A on March 11, 2016.
66 Interview on August 24, 2016.
67 Interview on January 15, 2016.
68 Interview A on March 11, 2016.
69 Interview on August 26, 2016.
70 Interview on March 6, 2016.
71 Interview on August 24, 2016.
72 Interview B on August 23, 2016.
73 Interview A on March 11, 2016.
74 Interview on August 16, 2016.
75 Interview B on August 23, 2016.
76 Interview on August 16, 2016.
77 Interview on March 6, 2016.

References

Altieri, M.A.; Funes-Monzote, F.R.; Petersen, P. 2012. Agroecologically efficient agricultural systems for smallholder farmers. Contributions to food sovereignty. *Agron. Sustain. Dev.* 32, No. 1, 1–13. DOI: 10.1007/s13593-011-0065-6.

Anderson, M.; Pienaar, K. 2003. *Municipal commonage.* Cape Town: Programme for Land and Agrarian Studies, School of Government, University of the Western Cape (Evaluating land and agrarian reform in South Africa - An occasional paper series).

Bayat, A. 2000. From 'dangerous classes' to quiet rebels'. Politics of the urban subaltern in the Global South. *International Sociology* 15, No. 3, 533–557. DOI: 10.1177/026858000015003005.

Borras, S.M.; Franco, J.C.; Suárez, S.M. 2015. Land and food sovereignty. *Third World Quarterly* 36, No. 3, 600–617. DOI: 10.1080/01436597.2015.1029225.

Claasen, N.; van der Hoeven, M.; Covic, N. 2016. Food environments, health and nutrition in South Africa. Mapping the research and policy terrain. *PLAAS Working Paper* 34, No. Cape Town.

CNdV Africa 2016. Thembalethu Precinct Plan. Urban Upgrade Report. http://www.george.org.za/sites/default/files/documents/Thembalethu%20Precinct%20Plan%20Report%20-%20Final%20-%20March%202016.compressed.pdf, Accessed on 27/03/21.

Department of Land Affairs 2006. Implementation plan for the proactive land acquisition strategy. Republic of South Africa. https://www.gov.za/sites/default/files/gcis_document/201409/impllandacquisition0.pdf, Accessed on 05/03/21.

Ferguson, R.S.; Lovell, S.T. 2014. Permaculture for agroecology. Design, movement, practice, and worldview. A review. *Agron. Sustain. Dev.* 34, No. 2, 251–274. DOI: 10.1007/s13593-013-0181-6.

George Herald 2012. Farmers protest for land. Published 06/07/2012. George.

George Herald 2014a. 52 American visitors explore George. Published 31/03/2014. George.

George Herald 2014b. Eve se vele rolle as vrywilliger. Published 24/03/2014. George.

George Herald 2015. Food gardeners help feed the city. Published 10/12/2015. George.

George Herald 2016. Kos en Fynbos values affirmed. Published 21/01/2016. George.

George Municipality 2016. Integrated development plan 2012–2017. 4th Review 2016/2017. George. http://www.george.gov.za/sites/default/files/documents/George%20Municipality%20Final%20IDP%20Reveiw%202016-17.pdf, Accessed on 04/04/2021.

Holt-Giménez, E.; van Lammeren, I. 2020. Can food as a commons advance food sovereignty? In: Vivero-Pol, J.L. et al. (Eds.): *Routledge handbook of food as a commons.* London: Routledge. 313–328.

Kerkvliet, B.J.T. 2009. Everyday politics in peasant societies (and ours). *The Journal of Peasant Studies* 36, No. 1, 227–243. DOI: 10.1080/03066150902820487.

LED 2013. *George consumer survey.* George: George Municipality.

LED 2016. *Project plan: Household food security project phase 1: March – June 2016.* George: George Municipality.

Lefebvre, H. 1971. *Everyday life in the modern world.* London: Allen Lane Penguin Press.

Lefebvre, H. 2002. Comments on a new state form. *Antipode* 33, No. 5, 769–782. DOI: 10.1111/1467-8330.00216.

Martínez-Torres, M.E.; Rosset, P.M. 2014. Diálogo de saberes in La Vía Campesina. Food sovereignty and agroecology. *The Journal of Peasant Studies* 41, No. 6, 979–997. DOI: 10.1080/03066150.2013.872632.

Nyéléni 2007. Synthesis Report Nyéléni 2007 Forum for Food Sovereignty. Forum for Food Sovereignty. nyeleni.org/IMG/pdf/31Mar2007NyeleniSynthesisReport-en.pdf, Accessed 09/01/21.

O'Brien, K.J. 1996. Rightful resistance. *World Pol.* 49, No. 1, 31–55.

O'Brien, K.J. 2013. Rightful resistance revisited. *Journal of Peasant Studies* 40, No. 6, 1051–1062. DOI: 10.1080/03066150.2013.821466.

O'Brien, K.J.; Li, L. 2006. *Rightful resistance in rural China.* Cambridge, New York: Cambridge University Press (Cambridge Studies in Contentious Politics).

Peyton, S.; Moseley, W.; Battersby, J. 2015. Implications of supermarket expansion on urban food security in Cape Town, South Africa. *African Geographical Review* 34, No. 1, 36–54. DOI: 10.1080/19376812.2014.1003307.

Purcell, M. 2014. Possible worlds. Henri Lefebvre and the right to the city. *Journal of Urban Affairs* 36, No. 1, 141–154. DOI: 10.1111/juaf.12034.

Roy, A. 2011. Slumdog cities. Rethinking Subaltern Urbanism. *Int J Urban Regional* 35, No. 2, 223–238. DOI: 10.1111/j.1468-2427.2011.01051.x.

Scott, J.C. 1985. *Weapons of the weak. Everyday forms of peasant resistance.* Princeton, N.J.: Yale University Press.

Skinner, C.; Haysom, G. 2017. The informal sector's role in food security. A missing link in policy debate. *Hungry Cities Partnership – Discussion Papers, No. 6.* http://hungrycities.net/wp-content/uploads/2017/03/HCP6.pdf, Accessed 11/01/21.

South African Cities Network 2014. George land of milk and honey? *SACN Research Report,* University of the Free State. http://www.sacities.net/wp-content/uploads/2015/10/George-report-final-author-tc-3.pdf, Accessed on 22/04/2021.

Western Cape Government Provincial Treasury 2018. Municipal economic review and outlook 2018. https://www.westerncape.gov.za/assets/departments/treasury/Documents/Research-and-Report/2018/2018_mero_revised.pdf, Accessed on 25/02/21.

Yamashita, L.; Robinson, D. 2016. Making visible the people who feed us. Educating for critical food literacy through multicultural texts. *Journal of Agriculture, Food Systems, and Community Development,* 269–281. DOI: 10.5304/jafscd.2016.062.011.

5   Politicising alternatives from below

This chapter departs from vital debates and pending questions on the wider impact and possible change community initiatives like KEF are able to create. Policy-makers, community workers, academics, and movements themselves are often confronted with the critical question: Are these so-called alternatives "from below" niches within the unequal agri-food landscape or are they able to propel wider changes beyond their community? It is thus the intention to provide answers and recommendations in the specific case study context and beyond. This discussion adds critical considerations evident in broader research on urban agriculture and food sovereignty in different regions and thus provides guidance on how to politicise efforts towards food sovereignty in the city and across the margins for diverse community initiatives, food movements, and their allies.

In general, the KEF initiative dismantles inequalities in the agri-food system and constructs alternatives for a socially and ecologically resilient food system, which frame the groups' political work. More precisely, the initiative targets local-ised food production, healthy nutrition, retail structures, land access as well as soli-darity networks, which relate to a range of urban pressures and racial segregation in the past. Government departments and the municipality are still overburdened with land requests for food production in the city. The members' experiences also show that the retail landscape has changed tremendously over the last years. Informal shops and traders have become under increasing pressure from the prolif-eration of supermarkets. However, both still continue to play important roles as food source for consumers and point of sale for some food producers. This chapter considers these case-specific responses from below in the context of the larger food sovereignty and urban agriculture debate. The work of the group has shone a light on realities, which have not been dominant in the prevailing food sovereignty dis-course, for instance the role of the informal sector in urban food provision and food producers' deprived realities at the urban outskirts. The latter shows parallels to struggles of small-scale farmers in the countryside and thus calls for food sover-eignty dialogues beyond the rural–urban divide. The chapter emphasises the importance of further politicisation including strategising and alliance building beyond isolated niches. For this purpose, the chapter is divided into three parts. First, building on the previous chapters, the political dimension of KEF's work and inherent diverse rights to the city are uncovered. The introduction of these rights is guided by the overall question of what can be taken from the experience of KEF

DOI: 10.4324/9781003182634-5

to further define and strengthen food sovereignty's urban dimension and community interventions in George and elsewhere. Such local claims and everyday politics for a more equal agri-food system require an entry point to local politics, as well as solid and participatory urban food governance. Departing from the politicisation of KEF, further suggestions are made on how effort towards food sovereignty could be scaled-up. Second, considering the felt divide between rural and urban realities in the food sovereignty discourse and movements, a stronger emphasis on the uniting elements, specifically similar experiences of those residing on the margins, is suggested to reap wider impacts and to create strong alliances. Beyond highlighting opportunities, this part comprises a critical discussion on the possible restrictions of community-based initiatives. The final part of this chapter gives nuance to discussed transformative attempts in the context of the broader social system. Here, the critical urban food perspective and specifically Eric Olin Wright's (2017, 2019) anti-capitalist strategies are used to differentiate and contrast food sovereignty efforts and everyday politics. A critical review of related considerations in the literature and a fresh perspective inspired by Wright's take provides helpful guidance on how a more democratic agri-food system can be initiated. This chapter clearly highlights key take-away messages which are considered as helpful for the proliferation of transformative efforts of urban food producers and their allies in other parts of the world.

Uncovering political dimensions and rights to the city

In the previous chapter, the analytical steps of "exposing" and "proposing", in connection with the triad of space, revealed the challenges and limitations with which the KEF initiative is confronted. Simultaneously, active responses and ways of adapting and overcoming these situations were outlined. These struggles in everyday life point to variegated forms of resistance ranging from overt, rightful resistance against planned government evictions at the fringes of Thembalethu, to subtle practices of grocery purchase in local shops to avoid commercialised supermarkets. Building from these lived experiences, this section sets out to explore the political dimension – the step of "politicising", embodied in expressions of alternative and everyday politics (Eizenberg 2012, 765; Kerkvliet 2009, 229; Marcuse 2009, 194). The open definition of political acting under the right to the city serves as an ideal starting point to define food sovereignty-related rights to the city. It is key to contrast the experiences of the movement with responses "from above" in the local urban policy context. To better illuminate the starting of the initiative's political work, a brief overview of failures on part of the local government is presented first.

The work of KEF reveals the lack of an adequate institutional response to wider inequalities in the urban agri-food system. This includes for instance often-interrelated problems of food insecurity, nutritional deficiencies, and unemployment. The municipality of George and the regional Department of Agriculture have mainly targeted food poverty through the promotion of urban agriculture. In general, so-called quick win projects remain isolated responses and mirror the lacunae of broader and more systemic interventions against malnutrition. Public policies and municipal units do not explicitly refer to or contain food and nutrition interventions. Hence, NGOs and CSOs, for example community soup kitchens, continue to play an essential role in providing basic and healthy meals. Material and educational support (i.e. training) suffers from many shortcomings. Both institutions have failed thus far to provide adequate market access and seem overwhelmed by urban food producers' land demands. At the same time, the governance processes and interventions of the George Municipality and the (regional) Western Cape government proved to be very hierarchical (i.e. top-down). Thus, demands and development efforts introduced by the civil society are often not considered in public policies.

The dynamic and ever-growing commercialised retail sector implies that the government is a facilitator of neoliberal developments and refrains from regulations, for instance regarding price policies or the products on offer. In response, the informal sector appears to provide (emergency) solutions, for instance in food provision as well as market and land access for food producers. This "economy from below" pops up in all areas where the formal and official have proved inadequate, too expensive, out of reach. The weak integration of the informal sector and proliferation of this so-called shadow economy remains a big challenge all over South Africa (Skinner and Haysom 2017; Skinner 2018). Formalisation and further support of these informal structures are needed. Thus, municipalities are urgently required to rezone market areas and provide appropriate infrastructure in close exchange with the local communities. At the same time, this kind of "localisation" of the food system bears some challenges. Local food systems including retail structures are often idealised as inherently "good" – as just, ecologically sustainable, and democratic. Specific conditions of food production, preparation, or retail are not considered. In this regard, Born and Purcell refer to the so-called local trap (2006). They argue that scale is socially produced: "Localizing food systems, therefore, does not lead inherently to greater sustainability or to any other goal. It leads to wherever those it empowers want it to lead" (ibid., 196). For instance, it is important to think critically about the products on offer; the promotion of sugary soft drinks might not be an ideal choice. Related careful considerations are required in urban food system governance.

In both the George Municipality and the Western Cape Government, governance processes and structure are very hierarchical. Hence, participatory planning and explicit entry points for active citizens are barely existing. Personal connections with officials seem to be the only way to get certain demands or needs addressed, as exemplified by the strong connection of the movement to the previous manager of the LED. While being open to citizen-driven suggestions and the needs of the communities, the manager was also restricted in actions by the official municipal body and specific development plans. To sustain the efforts of KEF, an intense dialogue between policy-makers and communities must be established, which opens up space for truly participatory urban food policies. This may then comprise more adequate promotion of local, small-scale food production, improvement of the so-called shadow economy as well as wider availability and accessibility of healthy produce. In this respect, it is also essential for the authorities to engage intensively with the root causes of food insecurity and poverty. While there is much to be done here to establish citizen-oriented urban food governance, a number of suggestions already exist around the world for the integration of food sovereignty into urban policies (Desmarais, Clays, and Trauger 2017; García-Sempere et al. 2018, 398 ff.; Kay et al. 2018).

Beyond these shortcomings on part of the government, KEF fights on a number of fronts against "food dis-ability", which Tornaghi describes as "the absence of meaningful food experiences coupled with (and made possible by) the agro-food industry and corporate global supermarkets in its double grip on land control and the commodification of the food experience" (2017, 791). In this context, urban food production can be seen as politics. For KEF, the everyday politics are marked by the indispensability of certain practices, such as food production, land demands, and occupations, connections to the informal market, sharing of food, knowledge, and seeds. For instance, the motivation to grow food can be traced back to the manifold problems of insufficient food access, little financial means, health concerns, and the critical perception of the food provided by the market. It also includes limited governmental interventions and support, demonstrated for example by restricted land access and selling opportunities. The creation and conservation of knowledge about gardening is another intervention which has grown out of the need for (partial) self-provision as well as the absence of comprehensive knowledge provision and appreciation from the authorities. For instance, gardening and cooking skills are not part of the official curricula in schools. In this respect, Marcuse reminds us that the right to the city is "a cry out of necessity and a demand for something more" (2009, 190). In sum, daily practices of food production and consumption are politicised as they can be unpacked to reveal diverse forms of neglect. Through these daily practices, the movement creates visibility and a mode of operation, and explores what might be possible politically. Some of its practices are deeply incorporated in everyday life and thus remain rather hidden, like using medicinal plants. In contrast, other actions seek public attention, followers, supporters, and recognition by governmental bodies, for instance, land occupations, gardening demonstrations, and spreading the word through the newspaper (cf. O'Brien 1996, 34). In this way, KEF inspires many people to question the prevailing food landscape.

The movement incorporates different directions and targets of political action, which can be framed at two levels – the mostly local, and the wider national or even global level. The first targets local government – the municipality and different departments, for instance regarding the land debate, the prevailing retail structures, and development projects. The second level expands on the first and is an overall critique of the (national and global) commercialised food system, which is not directly in the hands of local (political) actors. Of course, the two levels cannot be neatly divided, and there are overlaps. In this sense, members pointed for instance to long-distance transportation of food and the often-related low quality, or the usage of GM crops. This is intertwined with the movement's critique of large-scale agriculture and its interest in self-sufficiency, as illustrated by a statement of one of the members:

> if everybody can sustain themselves with fruit or even chickens. [...] in a small sustainable way putting pressure on the huge agricultural large-scale agricultural farming that are actually the ones which are causing the destruction of our biodiversity. [...] we have our own food and vegetables and we pick them.[1]

It is important to acknowledge that the initiative does not use the food sovereignty rhetoric explicitly. Rather, its daily work in digging the soil and inherent resistances are expressions of food sovereignty in practice. While some members have heard about food sovereignty, it is not the dominant banner under which they have been mobilising and it is not an explicit part of the vision. However, the work of KEF builds the linkage to food sovereignty in practice and in everyday life, through the movement's focus on local ecological production, solidarity within and beyond communities, taking a critical stance towards government projects and the commercialised retail landscape, as well as their demand for land. In this respect, one can argue that they "walked the walk" in engaging critically with the prevailing agri-food system and wider inequalities in their city, which is similar to Figueroa's experience with the Chicago-based Healthy Food Hub. Based in a black community, the food hub collectively purchases food items. In this way, members get "good food for less money", strengthen cultural ties, and create "resilient infrastructures for community survival and independence from the economic forces that have enslaved, exploited and ultimately left them behind", which implies efforts towards food sovereignty (Figueroa 2015, 500 and 510). However, Figueroa also highlights that the members did not explicitly "talk the [food sovereignty] talk" (ibid. 510).

In general, food initiatives' local actions against exclusionary dynamics are an invitation to approach food sovereignty in places where it is less expected. Similar to Visser et al., one can assert that a kind of food sovereignty does exist in George; as it "thrives without any organisation that could formulate outspoken discourse or coordinate actions" (2015, 514). Their term "quiet food sovereignty" – food sovereignty without ties to a bigger movement – covers many aspects of the KEF experience. For instance, when asking members about organisations and campaigns advocating for food sovereignty at the national level, they could not answer. However, one very engaged member reflected: "You know, I would actually be very interested. I've been invited to the Slow Food celebrations in Italy. […] It's actually to see whether there is enough resources to go further with this thing [KEF]".[2] She assumes their localised initiative must first grow stronger before connecting with bigger movements. Several KEF members had a rough understanding of what food sovereignty implies and were strongly in favour of it: "localised things, localised production, localised buying power, localised everything. Support the local economy. […] I think, we are undermining the true side of food sovereignty […] the skills of people, and tiny stuff and shops. It's gone".[3] Linkages with organisations and groups working directly under the banner of food sovereignty could help the initiative to further polticise their work. For instance, the NGO Surplus People Project, which engages for instance in urban land redistribution, seems to a helpful contact.

The movement has been manoeuvring on different fronts but lacks a clear and precise strategy. It is more of a silent agreement between the members. Again, nuances of O'Brien's work can be found here, as individuals "come to appreciate common interests, develop oppositional consciousness, and become collective actors in the course of struggle" (1996, 34). However, he also highlights that they must develop tactics and strategies to be more effective. While this might remain a future task of KEF, the lived experiences of the members outlined in the previous chapter already point to different claims and rights to the city. It is key to highlight here that Lefebvre and several scholars have referred to multiple rights under the banner of the right to the city (Heynen 2010; Marcuse 2009; Tornaghi 2017). These rights if clearly defined could be essential elements of a future strategy and essential to "grow food sovereignty" in the city. Following the lead of the food sovereignty concept and pillars, these claims of the KEF initiative could be summarised as several interrelated rights to the city (cf. Tornaghi 2017, 792):

(1) *Right to access quality and locally grown food,*
    This demand refers to the problem of malnutrition, comprising an increasing availability of low quality (e.g. processed and poor in nutrients and vitamins) and cheap food (see Chapter 4, section "Local access to nutritious food").

(2) *Right to control own diets,*
    This demand is related to the previous right and points to overcoming restricted food choices (e.g. because of prices, limited time, or availability). Thus, it is the intention to return food choices on what (kind of food), where (food is sourced), and when (food is eaten) to the people (see Chapter 4, section "Local access to nutritious food").

(3) *Right to have a determining influence on local market environments,*
    Relating to the previous claim, this demand seeks improved choices in where to purchase food and where to sell produce (market access e.g. beyond informal food stalls) (see Chapter 4, section "Valuing food producers").

(4) *Right to have a determining influence on governmental community interventions,*
    This refers to the demands of the communities being understood and taken into consideration in governmental development interventions; both were lacking in previous attempts to improve food security and to provide income opportunities (see Chapter 4, section "Valuing food producers").

(5) *Right to access cultivable land,*
    This refers to the demands for land access for community gardens and farming. In particular, access to farmland in Thembalethu and integration into governmental development projects remained a continuous bargaining process, with land occupations being the sad reality (see Chapter 4, section "Access to land").

(6) *Right to an enabling knowledge environment,*
    This demand covers access to knowledge, e.g. for ecological food production and healthy cooking practices, and the expansion of traditional knowledge, e.g. conservation and application of traditional knowledge (see Chapter 4, section "Community knowledge and skills").

(7) *Right to environmental conservation,*
    Enshrined in practices of environmental protection and biodiversity creation, these demands seek further acceptance of and support for these activities and the reduction of harmful interventions such as dumping see Chapter 4, section "Biodiversity, permaculture, and connection to nature".

(8) *Right to solidarity,*
    Stepping beyond isolated interventions in neighbourhoods and communities, this is the demand to build stronger ties and cooperation between different actors in the city, beyond its boundaries, and in the agri-food system.

These different rights are defined here in their broadest sense to comprehensively cover the needs of the movement and to inspire consideration and application elsewhere beyond the specific case. Possibly more precise formulations could be an advantage in prospective strategies and further interventions on the ground.

Considering these inherent claims and rights shows parallels to O'Brien's notion of rightful resistance. Although he refers to rights which have been implemented, he shows that successful rights advocacy "sparked new dialogues, strengthened new ways of thinking, strengthened social identities, posed troubling questions, and spurred desire for political change" (1996, 52–53). KEF already engages in these tasks by reaching out to the public in diverse ways, e.g. gardening demonstrations and newspaper articles.

Beyond the defined demands, one can suggest that KEF is claiming further rights, for instance to (secure) access to the formal labour market. Although many of the movement's members are struggling in terms of income, livelihoods are highly diverse, and it is difficult to frame a demand which meets the needs of everyone (e.g. some might demand higher social grants or pensions, others higher salaries, or formal contracts). More importantly, these needs were not actively addressed by the members. While discussions revealed challenges related to these needs, they did not amount to specific claims or complaints. Nevertheless, it is essential to keep these labour market restrictions on the radar.

In this context of marginalisation, it is important to keep in mind that some actions might backfire and achieve the opposite of what is intended. For instance, as self-help increases, the state tends to take a step back and pay less attention to the needs of the people. On the other hand, the movement could also be at risk of co-option by the state and the neoliberal urban project (McClintock 2014; Tornaghi 2017). While some food producers complain about the lack of selling opportunities for their produce, one might wonder whether the desired end of their demands could be fixed contracts with large supermarket chains. In this sense, they would become part of the market-driven imperatives that they are currently criticising. Considerations like these might be helpful to KEF in weighing its ideals and their broader feasibility.

Ultimately, the rights drafted here originate in joint experiences of deprived dwellers which make everyday resistance and alternatives necessary to secure life quality beyond basic survival in the city. This can be termed active citizenship and seen as a proposal for participation in decision-making processes beyond a passive role in government programmes. Broader rights to appropriate space and to participate as part of the right to the city are mirrored (Purcell 2002, 102). Members claim the right to play a central role in decisions that contribute to the production of space. They claim the right to appropriate space by physically using and occupying it according to their needs. In this sense, it is important to keep in mind that moving towards the right to the city is not only enshrined in inhabiting urban space, "it is also a matter of attacking the wider processes and relations which generate forms of injustice in cities" (Iveson 2011, 250). Based in the deprived working class, the majority of the group is un(der)employed, relies on social grants, pensions, and on the food produced in their gardens. A smaller group of members is employed in jobs related to the initiative such as social work. Although the engagement in self-provisioning is more essential for the first group, the movement on the whole is conscious of common complaints and has a shared awareness of local historical particularities, e.g. the apartheid legacy. For instance, members with different backgrounds – black, coloured and white, affluent and unemployed, women and men, young and old – initiated a joint struggle to claim the land of the buffer zone and use it as common ground. Blocked by several government departments, the empty zone remained a sad reminder of the racial divisions under the apartheid regime. Marx would certainly doubt class-consciousness (i.e. class in itself) and thus the formation of a social class opposed to the ruling class (i.e. class for itself). Apparently, the experience of KEF, with its diverse membership, requires more openness and consideration of common elements as the uniting and nurturing factor in claiming rights to the city. Thus far, different backgrounds have not been a dividing factor in the initiative. Slight differences and roots in different communities propel strong interactions, allow interventions on diverse fronts (e.g. occupying land, demanding shared resource sites, emphasising the needs of malnourished patients, abstaining from municipal donations) and contribute to a collective construction of identity as urban food producers. This chimes in with Wittman's concept of agrarian citizenship.

Certainly, mobilisation around issues of food and land may turn passive consumers into active citizens and construct a new identity strongly shaped by the cultivation of food. This can be linked to the notion of agrarian citizenship which comprises food sovereignty as its central message (Wittman 2009a, 2009b). According to Wittman, many deprived people, for instance "living on the margins of urban areas as day laborers, have never accomplished political enfranchisement […]. The transformation of their personal self-vision and relation to a larger social network, through the collective struggle for land, alters the politics of the possible and broadens horizons for action" (2009a, 129). Similar to the right to the city, she asserts that agrarian citizenship is not an assumed right; it is rather protectionism of social and ecological spaces as well as an accomplishment – reinventing identities as active citizens beyond the power of the state (Wittman 2009a, 129; 2009b, 822). Parallels can be found in the lived realities of many KEF members who barely experienced political enfranchisement under the apartheid regime, particularly referring to the squatter farmers at the fringes of Thembalethu. The KEF's network helps them to organise and strengthen their land demands. Similar considerations are also evident in O'Brien's work, who assumes that "growing rights and rules consciousness is the path along with growing citizenship rights" (2013, 1059). Wittman's ideas enrich the critical urban food perspective and offer important guidance to understand the identities and realities of food producers in cities, which might even refer to new peasant spaces, as per Jacobs (2018). This takes into account the ongoing construction and contestation of social identities (e.g. as worker/smallholder) (Wittman 2009a, 129). The case of KEF shows the nuances in the different identities of its members, for instance as part-time food producers and critical consumers. However, these identities are contested by various neoliberal dynamics, which restrict sufficient integration in the labour market, and limit land access or food choices. Agrarian citizenship implies challenging these discriminations. These experiences from the fringes of a city and intertwined urban-rural identity call for more flexibility in the food sovereignty discourse and movement.

Growing food sovereignty across the rural–urban divide

Recent work on food sovereignty is firmly rooted in, and emphasises, struggles in the countryside. It is stressed that rural and urban settings vary; producers and consumers differ in their connection to nature, demands, and needs. This is also referred to as the urban–rural dichotomy or rural–urban divide (e.g. García-Sempere et al. 2018). Others put more emphasis on the elements that connect urban and rural struggles, including agrarian, food, and environmental issues as well as livelihoods (Borras et al. 2018, 1227; Du Toit 2018, 1098; Hart and Sitas 2004). It is in this context that many scholars and food sovereignty activists articulate the need to unite struggles and facilitate cooperation.

Following this lead, the critical urban food perspective suggests to adjust the lens, beyond the assumption of a divide, and rather focus on the evident overlaps, which may facilitate joint interventions. The experience with KEF and variegated forms of resistance, the combination of diverse livelihood strategies, land struggles, and pressures in social reproduction, all point to the interlinkages and similarities between rural and urban settings. The roots in the countryside and in food production of many KEF members show how deprived realities in both settings influence each other.

Historically, inequalities in land access and ownership have been constant companions in South Africa since the early decades of the twentieth century. Particularly at the end of apartheid, land reform gained attention, and comprehensive land reform programmes were introduced in an attempt to correct historical distortions and redistribute land to smallholders (Cousins 2016). However, limited outcomes have been criticised in many ways and fuelled public protests (Hendricks, Ntsebeza, and Helliker 2013, 1; Kepe and Hall 2018, 134). More recently, land expropriation debates are reaching new heights with a ferocious populist turn (Hart 2019, 318–319). Partly to shore up the declining electoral support for the ANC, President Cyril Ramaphosa decided to explore the politically explosive issue of possible expropriation without compensation in land reform in 2019. Whether such a rhetoric might translate into actual reality, and if so, how and to what extent, remains to be seen at the time of writing. However this unfolds in the future, it is relevant to point out in relation to the argument being advanced in this book that the focus of the official land discourse has mainly been on rural areas. This is consistent with governmental reform attempts since 1994 (Walker 2007, 141). All this has influenced activist advocacy and academic research where political and academic engagement with the urban dimension of the land question is relatively limited (Hendricks, Ntsebeza, and Helliker 2013, 2; Jacobs 2018, 16; Joseph, Magni, and Maree 2015, 4). A rich body of literature deals with the historical and contemporary land problem (for instance, see the works of Henri Bernstein, Ben Cousins, Ruth Hall, and Sam Moyo).

Rural land inequality, enclosures, and commercialisation of agriculture have been fuelling in-country migration and are linked to South Africa's urbanisation, with 67% of the population living in cities (United Nations Population Division 2019). In cities, many people are engaged in informal labour and multiple income activities, which is also evident in George. The precariousness of these livelihood strategies has caused many to turn to urban farming. However, land access has remained an issue. As indicated earlier, about one-third of black Africans demand land for agricultural production; 34% of this demand is from urban areas (Aliber, Reitzes, and Roef 2006). These demands often take the form of land occupations; as Hall puts it: "Occupation of land, in rural or peri-urban areas, demonstrates a need expressed by people voting with their feet" (2009, 70). In the case of Thembalethu in George, farmers have been waiting for official land access for more than 20 years. Others have been struggling to farm the leased land efficiently. While many of them left deprived regions in the Eastern Cape, their roots in farming are still vital for survival in the city. Despite land reform interventions such as PLAS, squatter farmers have not benefited from the programme and have been continuously at risk of eviction. The story of these farmers mirrors the country's difficulties in engaging properly with the land question and related historical tensions. In addition, cities are still shaped by the former apartheid architecture and thus the country's history of racial segregation and exclusion, with a relatively high number of people living in informal dwellings with poor basic service provision. High and volatile food prices are a further motivation for self-provision. Hence, informal structures of food production are indispensable.

It has to be emphasised that land still plays an integral role in survival in both rural and urban areas. Beyond insecure land tenure systems and inadequate land access for the rural poor, scholars agree that urban and peri-urban land reform remains a challenge in South Africa (Du Toit 2018; Hendler 2015). Over the last two decades, the urban land question has become prominent in the context of ever-increasing housing demands (Hendricks and Pithouse 2013; Tapscott 2010, 266–267), but there has been a failure to engage adequately with increasing requests for farmland (Jacobs 2018). Despite the contestation in urban land use, Joseph, Magni, and Maree strongly emphasise the responsibility of South African cities to consider and balance diverse land use purposes, particularly in housing, urban agriculture, infrastructure, and open space (2015, 3 and 6). This kind of diversification in land use is challenged by the neoliberal urban land greed, related urban policies, and continued in-migration to cities. These experiences and similar struggles over land for agricultural purposes have become more widely recognised around the world, as illuminated for instance in the works of Chappell (2018) on Brazil, Bowness and Wittman (2020) on Canada, Gillespie (2016) on Ghana, and Purushothaman and Patil (2019) on India.

One key message of the critical urban food perspective is that people across urban and rural settings often share similar difficulties, and the problem is larger than the land debate per se. It is of overall importance in ensuring sustainable and sufficient livelihoods (Hendricks, Ntsebeza, and Helliker 2013, 3–4). Therefore, the decisive task remains to figure out how livelihoods can be made secure in the context of marginalisation, exclusion, precariousness, informality, jobless de-agrarianisation in urban and rural areas, particularly at the margins. In addition, the question arises: how mixed and multiple livelihood strategies can be rendered redundant by market and capitalist expansion globally.

Beyond the questions of land and livelihoods, it is important to grapple with the weak and neglected consumer dimension in the food sovereignty discourse which is often perceived to be an urban phenomenon as urban dwellers rely on food provision from outside. However, a more nuanced understanding of consumption and food provision is required. The consumer class also includes farm workers and rural inhabitants, who are not necessarily in a position to supply themselves year-round with self-grown food. In many areas around the world, complete self-sufficiency and subsistence have been declining, creating at least partial market dependency (e.g. Cousins et al. 2018; Sibhatu and Qaim 2017). Hence, it is essential to consider food provision as well as consumer needs and forces beyond the urban, at the periphery, in any transformative attempts regarding the commercialised agri-food system. The case of KEF illustrates that the marginalised and deprived are confronted with limited choices, including so-called imposed food, which does not necessarily reflect the needs of the consumer, is expensive and non-transparent in terms of origin, preparation, and ingredients. In this way, individuals have restricted control over their nutritional intake and might deal with health risks as a consequence. In addition, changing lifestyles and working patterns have made food and consumption abstract comprising purchase of low-quality food and rushed eating. This links to a critical consumer dimension of food, as embodied by KEF.

Several scholars point to the importance and positive experience of alliances between those struggling in and outside cities, for instance workers and peasants in different places around the world (e.g. Constable 2009; Jacobs 2013; Pye 2010). In Zimbabwe, for instance, the problem of inadequate land access led to a strong and powerful alliance between urban and rural forces in Zimbabwe's land occupation movement in the 1990s. Urban dwellers were hit hard by housing shortages and unemployment and thus decided to join the movement. However, taking a historical perspective, Mamdani is rather sceptical about this revolutionary potential, given the decline of smallholders. He refers to rural subjects and urban citizens and outlines that "the depth of resistance in South Africa was rooted in urban-based worker and student resistance, not in peasant revolt in the countryside" (Mamdani 1996, 29). However, a neat distinction between urban and rural struggles and livelihoods might become increasingly difficult. Du Toit emphasises that "The livelihood strategies and coping mechanisms of poor African households themselves also cut across the rural–urban divide" (2018, 1098). This connects with Zhan's and Scully's observations in South Africa that "the rise of precarious work and the continued crisis of rural economies have led to a double squeeze on spatially extended livelihoods" (2018, 1023). As observed by Shivji, "peasants and pastoralists, proletarian and semi-proletarians" often share difficulties in constructing their livelihoods (2017, 11; Borras et al. 2018, 1229). This is also exemplified in the case of KEF. Many members have been struggling to make a living both at the urban and the rural margins. Although broadly based in the deprived working class in black and coloured communities, livelihoods have been continuously constructed by many members, shifting between or combining land and labour-based strategies. Any kind of food production, ranging from small backyards to small farm plots, combined with conscious consumption, has been the defining momentum of the group and their identity. Here, it is essential to refer back to the earlier introduced notion of agrarian citizenship. It could even be suggested that the characteristics of a rural peasant and an urban consumer intersect. In this context, Borras et al. provide an important take-away message "contemporary movements [...] appear to be operating within the politics of 'intersectionality' – of class and other social identities" (2018, 1229).

While there is increasing overlap between issues and struggles of deprived food producers, it can nonetheless be said that well-established urban-rural alliances under the banner of food sovereignty are barely existing. In South Africa, food sovereignty campaigns and organisations advocating for food sovereignty on the national level do not necessarily have the means to reach out to a plethora of local-ised struggles. Beyond potential ties to larger movements and NGOs, it is essential to consider and to connect localised and rather hidden urban agriculture work and food sovereignty attempts on the ground. There might be other urban initiatives similar to KEF in South Africa, which would greatly benefit from a closer interac-tion with sympathisers, on both the national and regional level. A larger network would be helpful to step up efforts, to exchange, to strategise, and to mutually sup-port each other. This is certainly a message we can take beyond the South African context.

It is important to further stress the convergence of urban and rural struggles of deprived people, creating solidarity, sharing knowledge and experiences, neatly figuring out interlinkages and uniting elements. Hunger, un(der)employment or health issues are prevalent at the margins in both sites. These problems are often only treated through limited social grants by the state and the promotion of simpli-fied self-help interventions, as the municipality's provision of seeds and gardening tools in the case of KEF shows. It would certainly be possible to outline many more linkages and overlaps between rural and urban dwellers. In sum, then, it is essential to make sure that no one is left behind (in either rural or urban settings and the periphery as a whole) and thus to pay attention to shared and uniting elements.

Creating changes within and beyond the existing system

So far, the applied critical urban food perspective shone a light on food produc-ers' diverse forms of resistance and the creation of space in the dominant capitalist agri-food system. The task at hand is to give nuance to these transformative efforts and to question possible broader impacts beyond a local group. As McMichael emphasises: "the food sovereignty movement is not simply about peasants, or food; rather, it concerns the undemocratic architecture of the state system, its erosion of social and ecological stability, and its politically, economically and nutritionally impoverishing consequences" (2016, 654). In fact, this refers to complex and mul-tiple tasks of changing the existing capitalist system and calls for a critical reflection of the related role of community initiatives. In this context, Tornaghi asks, "under what conditions can urban agriculture escape its marginality and contribute to reimagining, reshaping and radically changing the food system, and in doing so liberate us – at least partially – from the absolute capitalist control over a funda-mental sphere of social reproduction?" (2017, 791). Against the background of the dominant capitalist agri-food system and its negative impacts, what kinds of change are urban food producers able to create? A similar question could be asked for food sovereignty attempts. Are these initiatives "niches" within the prevailing system? Or have they already managed to replace current structures? Kaika for instance considers this kind of initiatives as "immunologies" – ways to alleviate the symptoms of the current system without changing it (2017, 98). Certainly, these questions provide decisive impulses when seeking to advance food sovereignty and urban agriculture movements. The following part of this book uses Erik Olin Wright's notion of "eroding capitalism" (2017; 2019) to distinguish between different transformative endeavours introduced by initiatives like KEF, related restricting factors and to explore what is needed to achieve broader changes. Wright states that "[it] may simply be impossible to have a coherent strategy for the emancipatory transformation of something as complex as a social system" (2019, 10). Departing from systemic challenges, Wright sketches four main strategies for anti-capitalist struggles: smashing capitalism, taming capitalism, resisting capitalism, and escaping capitalism (2017).[4] Table 5.1 provides an overview of these strategies, which are further explained in the following section.

Combing these different strategies is perceived as essential to erode the prevailing system. Interestingly, these political strategies are mirrored in different ways in the KEF initiative as well as efforts of urban agriculture and food sovereignty activists. Mapping and structuring these kinds of political work provides useful insights and guidance for those seeking to reap broader changes beyond communities.

The civil society-rooted strategy of "resisting capitalism", especially, broadly frames KEF's work. According to Wright, this strategy is applied by "many grassroots activists of various sorts" and "may not be able to transform capitalism, but we can defend ourselves from its harms by causing trouble, protesting" (2019, 50). His examples briefly refer to environmental protests or consumer boycotts and therefore include resistance in everyday life. Resisting capitalism implies neutralising its harms and thus withholding efforts to benefit the capitalist class. For example, KEF has a strong preference for locally produced food and is critical of the dominance of corporate power in the retail sector. Building on community solidarity, it has strengthened informal food sharing practices. In addition, discussion on land occupations and rejected land requests reflect a critique of the government's profit-oriented land policies and unfulfilled promises of land redistribution. KEF also exemplifies some elements of Wright's "escaping capitalism" strategy in its attempts to strengthen self-sufficiency while still relying on certain grocery purchases. This strategy is not anti-capitalist as is it refers to ways of living within the system (Wright 2017, 16). Sage, Kropp, and Antoni-Komar mention similar

Table 5.1 Typology of anti-capitalist strategies

	Neutralising harms		Transcending structures
The state	Taming	Eroding	Smashing
The civil society	Resisting	capitalism	Escaping

(Adapted from Wright 2017, 16; 2019, 59)

micro-alternatives, for example resilient "small territorial units (the city- region)" as drivers for a larger transformation and "system redesign" (2020, 9). Clearly, stepping out of an "insular existence" is a decisive factor for wider changes.

Turning back to the right to city, many proponents would encourage the KEF movement and citizens' attempts to change the city in a truly revolutionary way. In its most radical sense, the right to the city calls for an overhaul of society, a revolution to reclaim the city – a citizen-driven movement which makes the state obsolete and builds autogestion (self-management) (Lefebvre 2002, 779–780; Purcell 2014, 150). In this sense, Wright speculates that scholars imagine "a new world from the ashes of the old"; this is what he frames as the strategy of "smashing capitalism" (2017, 10). It seems impossible for small-scale, alternative food producers to initiate such a rupture.

Against this background, it is important to consider varying conceptualisations and differentiations of urban agriculture when questioning its potential impacts. McClintock gives an overview of different types of urban agriculture and places them on a spectrum comprising neoliberal, reformist, progressive, and radical manifestations (2014, 160). The latter two manifestations, the progressive and radical, have parallels with the food sovereignty discourse in opposing the industrial agrifood system, controlling the means of production, challenging capitalism/oppression, and reclaiming the commons (McClintock 2014, 153–154). Holt-Giménez and Shattuck suggest a similar differentiation of food movements and particularly highlight the radical politics required in the struggle for food sovereignty, including redistributive land reform and community rights to water and seed (2011, 117-118). While some authors tend to argue that only radical movements are contributing to and creating food sovereignty (amongst others Holt-Giménez and Shattuck 2011; Sbicca 2014), this book follows McClintock's suggestion for a more flexible categorisation and a more sophisticated approach to understanding and appreciating urban agriculture's roots, impacts and related heterogeneity (2014, 148). This flexibility helps to comprehend urban agriculture's diverse potentials to change and transform the food system under the banner of food sovereignty.

The critical urban food perspective mirrors this kind of openness regarding the motivations and political claims of urban food producers but also in the way food sovereignty is approached. In general, the concept of food sovereignty might lead to two varying destinations. First, many proponents are entertaining food sovereignty's rather radical notion and might envision a revitalised peasant agriculture in a post-capitalist socio-economic model. This implies a *change of the (capitalist) system* and thereby the contemporary food system, which is, as Bernstein puts it, a historical world ambition (2014, 1031; Holt-Giménez and Shattuck 2011). In Wright's typology of anti-capitalist strategies, a rupture in the existing system is referred to as "smashing capitalism". He also refers to the difficulties of this strategy and its rare successes in the past (2017, 10–11). The second, less radical trajectory of food sovereignty refers to creating alternatives and spaces of contestation as a *change within the system*. In many cases, the struggle for food sovereignty is closely related to the goal of re-establishing the connection between peasants, the market and the state, rather than creating an alternative development model (Edelman 2014, 970; Henderson 2016, 51). In this vein, food sovereignty is constructed as a progressive political discourse that inspires and guides peasant struggles (Henderson 2016, 52). This relates with Scott's and O'Brien's forms of resistance. According to O'Brien, "like everyday resistance rightful resistance is a within-system form of contention, in the reform, not the revolution, paradigm" (2013, 1058).

Moreover, there is a tendency to differentiate radical and progressive motivations for food sovereignty along the north–south divide (Holt-Giménez and Shattuck 2011). Typically, movements building upon peasant struggles and involved in transnational agrarian movements like La Vía Campesina are the rather radical front in the south (Borras, Edelman, and Kay 2008, 188; Edelman and Borras 2016, 155). In contrast, the initiatives emerging in the north are often considered as conscious urban consumption movements.

While these categorisations build upon various experiences, this book's take is in line with McClintock: it is essential to go beyond these established frames and adjust our critical perspective accordingly. Scott for example highlights that: "Most of the political life of subordinate groups is to be found neither in the overt collective defiance of powerholders nor in complete hegemonic compliance, but in the vast territory between these two polar opposites" (1985, 136). Diverse layers of political action, such as everyday forms of resistance or overt protests, could be essential elements for change and could fit the strategy of resisting capitalism or even escaping capitalism in Wright's typology. Abstaining from strict characterisation via a relational perspective allows to identify the particularities of a specific case, the agenda of those involved and the achievements such movements are able to reap. Thus, exploring the lived experiences of urban producers, as emphasised in the critical urban food perspective, helps to avoid simplistic impact assessments of urban agriculture initiatives.

Wright's typology of anti-capitalist strategies is ideally suited to further nuance KEF's work. He emphasises, "What we need is an understanding of the anti-capitalist strategies that avoids both the false optimism of wishful thinking and the disabling pessimism that emancipatory social transformation is beyond strategic reach" (2019, 38). The proposed combination of diverse strategies for a gradual change and multi-layered erosion cannot be done by one social and political movement alone. In this sense, the diversity of breaks created by variegated urban agriculture and food sovereignty movements could be an initial element of eroding the corporate agri-food system and thus challenging capitalism.

Similarly and in an optimistic tone, Bayat and Biekart emphasise that, "the neoliberal city is not a one-way street where capital reigns exclusively. Constraints notwithstanding, there are also opportunities for new actors and constituencies to make their mark on the configuration of urban life" (2009, 823). Hence, the interventions introduced by such initiatives might open up windows of opportunity Following this lead and the theoretical implications of the right to the city, Marcuse points to "sectors of everyday life that are free of capitalist forms, operating within the capitalist system but not of it [...] those are the sectors of the economy and of daily life that are not operated on the profit system, that [...] rely on solidarity, humanity" (2009, 195).

As illuminated with the critical urban food perspective, the case of KEF thus illustrates how sharing networks contribute to partial independence of the food system in place and strengthen solidarity. These daily actions dismantle inequalities and create alternatives. So far, the initiative has resisted co-option attempts from governmental development interventions. Certainly, some of the group's interventions are more radical than others, for instance the land occupations, but on the whole, the movement has been able to initiate change and critical debates on diverse fronts.

Wright urges further intertwined actions in "taming capitalism" from above. KEF's actions are located in the sphere of civil society. However, the work of food sovereignty organisations and campaigns introduced in Chapter 2 could play an integral role in initiating further civil society-led strategies in interaction with the state. For instance, the South African Food Sovereignty Campaign (SAFSC) engages in "resisting capitalism" from below, e.g. through protests against bread corporations (South African Food Sovereignty Campaign 2015). Beyond that, the SAFSC and the German Friedrich Ebert Stiftung published the People's Food Sovereignty Act in 2018, which they seek to advance further, for incorporation into government policies. In addition, the NGO Surplus People Project fought several court cases for land access for smallholders in the Western and Northern Cape. Efforts like these illustrate the use of legal frameworks and might shape state policies. This chimes with the strategy of "taming capitalism" in using "state power to neutralize the harms of capitalism" (Wright 2017, 13). Eroding the dominant food system requires a diversity of continuous resistances and alternatives. Referring to food sovereignty, Iles and Montenegro de Wit suggest: "Movements need [...] to be ready to collaborate with – and as importantly, to resist – actors and processes at compatible scales. One cannot adopt a fixed, small-scale approach to confront such a flexible, "many-headed beast" as capitalist agriculture" (2015, 487). If firmly articulated by civil society, advanced by different actors including the state and the market, and targeted from different angles, exclusionary agrarian and food structures could be challenged (Borras, Franco, and Suárez 2015, 613).

As highlighted earlier, alliances, cooperation, and partnerships on eye-level are key to advance grassroots' demands and putting these into practice. So far, KEF has mostly introduced and shaped alternative politics in a lonesome fight. Cooperation with other local actors is mainly characterised through material support or training. Partnerships with other initiatives working under the banner of food sovereignty could be a further opportunity to position and politicise interventions more strongly. Local and national actors who engage strongly and visibly politically, for instance by challenging existing land distribution, could be beneficial to reap successes, and expand demands. In addition, the notion of scholar-activism gained increasing attention in struggles for food sovereignty, food justice, and agroecology (cf. Edelman 2009; Montengro de Wit et al. 2021). Beyond taking sides, academics in different parts of the world seek for cooperation, intend to make grassroots people's voices heard, create awareness of local troubles, and path alternatives.

Concluding remarks

This chapter illustrates the inherent political work of the KEF initiative. Beyond the specific case, implications for food sovereignty construction in the city, across the rural–urban divide, and wider transformations of the capitalist system are sketched out. The everyday practices and activist work of KEF in close relation to the food sovereignty pillars inspired eight rights to the city. These rights are to be understood as possible guidance to further intensify and strategise such important grassroots work. Hopefully, the proposed rights might inspire critical food growers and their allies in other parts of the world. Ideally, they may want to know what lessons can be derived from the experience of KEF, its challenges, and successes.

In light of the increasing corporate consolidation in the global agri-food system, Sage, Kropp, and Antoni-Komar emphasise that a so-called second generation of food movements, which is not primarily concerned with local struggles, is becoming more visible globally (2020). This ties in with the experience of the KEF initiative in confronting the prevailing exclusionary agri-food landscape through diverse forms of resistance. Although these actions are rather localised, members' perceptions show a critical reflection of prevailing exclusionary structures in the agri-food system. This is the starting point of the food sovereignty discourse, which is basically concerned with issues of democracy, local people's rights in self-determination, and citizenship (Chappell 2018, 61). Hence, it is possible to advance the urban front of food sovereignty in discourse and in practice through the work of engaged citizens on the ground. At the same time, existing and interlinked knowledge, activist, and academic work on food sovereignty as well as food justice and agroecology have the potential to further inspire and unite localised struggles (see Chapter 1). Particularly food sovereignty as globalised movement with localised roots can play an integral role here in propelling wider alliances. It helps to understand local struggles in broader debates on the agri-food system. Thus, food sovereignty can be a strong "pulling factor" for broader actions, for instance engagement with the political dimension of food insecurity, and related interaction with the state and market.

The KEF initiative's daily practices of food production reveal the construction of food sovereignty on different fronts (e.g. in food and land access) and the related political dimensions. This chapter shows that the pillars of food sovereignty acquire specific urban dimensions, in which the food environment, the informal sector, the land, and labour dimension and the construction of knowledge deserve particular attention. Food sovereignty values food producers: it is thus essential that food production in cities overcomes its status of a marginalised activity. KEF was born out of exclusion in the neoliberal city and in the prevailing food system. This exclusion particularly becomes visible through income and food poverty including related health concerns. As well as making these problems visible, interventions in food production represent an attempt to counteract these developments and create alternatives. Building from different interventions, a set of rights and claims for the prevailing agri-food system are framed. For instance, the "right to have a determining influence on local market environments" is sketched out, which refers to the demand for improved choices in where to purchase food and where to sell produce. Despite the village-like character of several neighbourhoods, close ties to the city centre shape the food retail system and render local, informal food purchase structures redundant in this smaller, so-called secondary city. Members of the initiative seek to counteract these developments. The initiative's work calls for governmental responses to its subtle claims and diverse resistances. Participatory urban food policies and urban food governance are very much needed in George. This might include support and integration of the informal sector which plays such an integral role in food provision but also in the sale of produce. Overall, their work and derived rights to the city provide an important contribution to the urban dimension of food sovereignty and show important parallels to rural realities in terms of food provision, land access, and livelihood creation. These parallels urgently require further attention in the food sovereignty discourse. The empirical findings revealed that people with different backgrounds come together and synthesise politically in their daily interventions. They mutually construct identities of which food production is the key element which chimes with Wittman's notion of agrarian citizenship.

Despite its attempts at change, KEF suffers from limitations, tensions, and precariousness. It is located within the cracks of the system, where individuals are not fully integrated in the labour market and struggle with insecure or limited land access or food insecurity. Thus, an inherent desperation drives the engagement with the initiative. In this way, idealism is not always coherent and consistent. It is more of an unspoken agreement between different members. For instance, some members admitted that natural pest control can sometimes be difficult and that they apply chemical solutions as they rely on a rich harvest. Others still buy some of their groceries in supermarkets. In this vein, an oversimplified glorification of the movement's work in localising the food system is not appropriate. Food from the region or the informal market is not necessarily of high quality or pesticide-free (Siebert and May 2016, 162–163). Born and Purcell refer to this problem as the "local trap" (2006). These kinds of internal contradictions do not necessarily limit the movement's daily work or impact, but a clear vision statement and strategy might help to strengthen loosely combined ideas and perceptions. Most statements regarding the vision of the initiative refer to a radical alternative in terms of self-sufficiency. Unsurprisingly, this imagination is not backed up by comprehensive practice. However, as illuminated in part three "Creating changes within and beyond the existing system", the combination of direct actions, outspoken interventions, bargaining with government representatives, public policy claims, and challenging of historical barriers might contribute to a broader change and an alternative vision of the city. This is demonstrated by the fact that KEF has been gaining ground in several communities, new people and supporters have joined, different actors have acknowledged the assets of the movement (e.g. university as well as the city), have listened to their claims, and sought – often only the most palatable – ways to support them. Guided by Wright's anti-capitalist strategies, it can be said in summary that the initiative's alternatives enshrine potential for a broader transformation of the neoliberal city and the commercialised agri-food system. While some of the aims might be termed "utopian impulses" (Gardiner 2004, 232), those continued and intertwined actions of exposing issues, proposing alternatives, and politicising are key to moving closer to implementing inhabitants' claims (Marcuse 2009). This is certainly one of the key take-away messages for food movements elsewhere.

In addition, the findings of the localised initiative show that their desires and claims would benefit from strong support from, and exchange with, other social movements and local initiatives working in similar fields and pursuing similar ideals.

Identities, livelihoods, and struggles of the members exemplify the intertwined nature of rural and urban struggles. Historical inequalities, dominant corporate power, and the tendency of the state to act in favour of the market, cause complex and diverse pressures on livelihoods and social reproduction across the broad rural–urban continuum. This calls for alliances with rural farmers, urban workers, consumer organisations, etc. in and beyond the local level. Cooperation with food sovereignty campaigns and NGOs introduced in Chapter 2 may help to advance these localised to grassroots struggles and so-called niche attempts. Food sovereignty movements and food sovereignty proponents could play an important role in uniting a range of issues and engaging with the interface of rural and urban struggles. In this way, more pronounced political demands could be created which relate to the clear role of food production in the initiative's identity (i.e. agrarian citizenship) and further application of the outlined anti-capitalist strategies would be possible.

The case of KEF shows that a reconfiguration of the urban food system requires specific demands of the people. Dialogue with and participation of the people is a vital step and might contribute to a comprehensive urban food governance with entry points for diverse actors. For instance, residents can engage in the creation and development of their neighbourhoods, which are, after all, meant for those who live there. As experts in these areas, they are in the best position to consider the hidden and informal aspects which tend to be overlooked by an external planner or municipal worker. In this context, it is essential for policy-makers to avoid looking at those target communities through a simplified deficit lens which ignores the knowledge and social capital already evident in these groups (e.g. Kaika 2017).

Notes

1 Interview on August 24, 2016.
2 Interview on August 16, 2016.
3 Interview on August 24, 2016.
4 More recently, in 2019, Wright defined "dismantling capitalism" as a separate strategy, which was previously partly covered in the strategy "smashing capitalism". For the sake of simplicity, this book does not differentiate between both. Wright himself emphasises, "both smashing and dismantling capitalism envision the ultimate possibility of replacing capitalism with a fundamentally different kind of economic structure, socialism" (2019, 44).

References

Aliber, M.; Reitzes, M.; Roef, M. 2006. *Assessing the alignment of South Africa's land reform policy to people's aspirations and expectations: A policy-based report based on a survey in three provinces*. Cape Town: HSRC Press.

Bayat, A.; Biekart, K. 2009. Cities of extremes. *Development and Change* 40, No. 5, 815–825. DOI: 10.1111/j.1467-7660.2009.01584.x.

Bernstein, H. 2014. Food sovereignty via the "peasant way": a sceptical view. *The Journal of Peasant Studies* 41, No. 6, 1031–1063. DOI: 10.1080/03066150.2013.852082.

Born, B.; Purcell, M. 2006. Avoiding the local trap. *Journal of Planning Education and Research* 26, No. 2, 195–207. DOI: 10.1177/0739456X06291389.

Borras, S.M.; Edelman, M.; Kay, C. 2008. Transnational agrarian movements: Origins and politics, Campaigns and Impact. *Journal of Agrarian Change* 8, No. 2 and 3, 169–204. DOI: 10.1111/j.1471-0366.2008.00167.x.

Borras, S.M.; Franco, J.C.; Suárez, S.M. 2015. Land and food sovereignty. *Third World Quarterly* 36, No. 3, 600–617. DOI: 10.1080/01436597.2015.1029225.

Borras, S.M.; Moreda, T.; Alonso-Fradejas, A.; Brent, Z.W. 2018. Converging social justice issues and movements. Implications for political actions and research. *Third World Quarterly* 39, No. 7, 1227–1246. DOI: 10.1080/01436597.2018.1491301.

Bowness, E. and Wittman, H. 2020: Bringing the city to the country? Responsibility, privilege and urban agrarianism in Metro Vancouver. *The Journal of Peasant Studies*, DOI: 10.1080/03066150.2020.1803842.

Chappell, M.J. 2018. *Beginning to end Hunger. Food and the environment in Belo Horizonte, Brazil, and beyond.* Oakland: University of California Press.

Constable, N. 2009. Migrant Workers and the many states of protest in Hong Kong. *Critical Asian Studies* 41, No. 1, 143–164. DOI: 10.1080/14672710802631202.

Cousins, B. 2016. Land reform in South Africa is failing. Can it be saved? *Transformation* 92, No. 1, 135–157.

Cousins, B.; Dubb, A.; Hornby, D.; Mtero, F. 2018. Social reproduction of "classes of labour" in the rural areas of South Africa. Contradictions and contestations. *The Journal of Peasant Studies* 45, No. 5-6, 1060–1085. DOI: 10.1080/03066150.2018.1482876.

Desmarais, A.A.; Clays, P.; Trauger, A. (Eds.) 2017. *Public policies for food sovereignty.* London: Routledge.

Du Toit, A. 2018. Without the blanket of the land. Agrarian change and biopolitics in post–Apartheid South Africa. *The Journal of Peasant Studies* 45, No. 5-6, 1086–1107. DOI: 10.1080/03066150.2018.1518320.

Edelman, M. 2009. Synergies and tensions between rural social movements and professional researchers. *The Journal of Peasant Studies* 36, No. 1, 245–265. DOI: 10.1080/03066150902820313.

Edelman, M. 2014. Food sovereignty. Forgotten genealogies and future regulatory challenges. *The Journal of Peasant Studies* 41, No. 6, 959–978. DOI: 10.1080/03066150.2013.876998.

Edelman, M.; Borras, S.M. 2016. *Political dynamics of transnational agrarian movements.* Rugby: Practical Action Publishing.

Eizenberg, E. 2012. Actually existing commons. Three moments of space of community gardens in New York City. *Antipode* 44, No. 3, 764–782. DOI: 10.1111/j.1467-8330.2011.00892.x

Figueroa, M. 2015. Food sovereignty in everyday life. Toward a people centered approach to food systems. *Globalizations* 12, No. 4, 498–512. DOI: 10.1080/14747731.2015.1005966.

García-Sempere, A.; Hidalgo, M.; Morales, H.; Ferguson, B.G.; Nazar-Beutelspacher, A.; Rosset, P. 2018. Urban transition toward food sovereignty. *Globalizations* 15, No. 3, 390–406. DOI: 10.1080/14747731.2018.1424285.

Gardiner, M. 2004. Everyday utopianism. Lefebvre and his critics. *Cultural Studies* 18, No. 2-3, 228–254. DOI: 10.1080/0950238042000203048.

Gillespie, T. 2016. Accumulation by urban dispossession. Struggles over urban space in Accra, Ghana. *Trans Inst Br Geogr* 41, No. 1, 66–77. DOI: 10.1111/tran.12105.

Hall, R. 2009. Land reform how and for whom? Land demand, targeting and acquisition. In Hall, R. (Ed.): *Another Countryside? Policy Options for Land and Agrarian Reform in South Africa.* Cape Town: Institute for Poverty, Land and Agrarian Studies, School of Government, University of the Western Cape, 63–92.

Hart, G. 2019. From authoritarian to left populism? Reframing debates. *South Atlantic Quarterly* 118, No. 2, 307–323. DOI: 10.1215/00382876-7381158.

Hart, G.; Sitas, A. 2004. Beyond the urban-rural divide: Linking land, labour, and livelihoods. *Transformation*, No. 56, 31–38.

Henderson, T.P. 2016. State–peasant movement relations and the politics of food sovereignty in Mexico and Ecuador. *The Journal of Peasant Studies* 44, No. 1, 33–55. DOI: 10.1080/03066150.2016.1236024.

Hendler, P. 2015. The right to the city: The planning and "unplanning" of urban space since 1913. In Cousins, B., Walker, C. (Eds.): *Land divided, land restored. Land reform in South Africa for the 21st century*. Auckland Park South Africa: Jacana Media.

Hendricks, F.; Ntsebeza, L.; Helliker, K. 2013. Land Questions in South Africa. In Hendricks, F., Ntsebeza, L., Helliker, K. (Eds.): *The promise of land. Undoing a century of dispossession in South Africa*. Auckland Park: Jacana Media, 1–26.

Hendricks, F.; Pithouse, R. 2013. Urban Land Questions in Contemporary South Africa. The Case of Cape Town. In Hendricks, F., Ntsebeza, L., Helliker, K. (Eds.): *The promise of land. Undoing a century of dispossession in South Africa*. Auckland Park: Jacana Media, 103–129.

Heynen, N. 2010. Cooking up non-violent civil-disobedient direct action for the hungry. "Food not bombs" and the resurgence of radical democracy in the US. *Urban Studies* 47, No. 6, 1225–1240. DOI: 10.1177/0042098009360223.

Holt-Giménez, E.; Shattuck, A. 2011. Food crises, food regimes and food movements: rumblings of reform or tides of transformation? *The Journal of Peasant Studies* 38, No. 1, 109–144. DOI: 10.1080/03066150.2010.538578.

Iles, A.; Montenegro de Wit, M. 2015. Sovereignty at what scale? An inquiry into multiple dimensions of food sovereignty. *Globalizations* 12, No. 4, 481–497. DOI: 10.1080/14747731.2014.957587.

Iveson, K. 2011. Social or spatial justice? Marcuse and Soja on the right to the city. *City* 15, No. 2, 250–259. DOI: 10.1080/13604813.2011.568723.

Jacobs, R. 2013. The radicalisation of the struggles of the food sovereignty movement in Africa. *La Via Campesina's Open Book: Celebrating 20 Years of Struggle and Hope*. https://viacampesina.org/en/wp-content/uploads/sites/2/2013/05/EN-11.pdf, Accessed on 12/10/2020.

Jacobs, R. 2018. An urban proletariat with peasant characteristics: Land occupations and livestock raising in South Africa, *The Journal of Peasant Studies* 45, No. 5-6, 884–903. DOI: 10.1080/03066150.2017.1312354.

Joseph, S.-L.; Magni, P.; Maree, G. 2015. Introduction. In South African Cities Network (Ed.): *The urban land paper series*. Volume 1. Braamfontein: South African Cities Network, 1–5.

Kaika, M. 2017. "Don't call me resilient again!": the New Urban Agenda as immunology … or … what happens when communities refuse to be vaccinated with "smart cities" and indicators. *Environment and Urbanization* 29, No. 1, 89–102. DOI: 10.1177/0956247816684763.

Kay, S.; Mattheisen, E.; McKeon, N.; De Meo, P.; Moragues Faus, A. 2018. Public Policies for Food Sovereignty. Think piece series Food for Thought No. 1. Transnational Institute. https://www.tni.org/files/publication-downloads/web_public_pol_food_sov.pdf, Accessed on 01/04/2021.

Kepe, T.; Hall, R. 2018. Land redistribution in South Africa. Towards decolonisation or recolonisation? *Politikon* 45, No. 1, 128–137. DOI: 10.1080/02589346.2018.1418218.

Kerkvliet, B.J.T. 2009. Everyday politics in peasant societies (and ours). *The Journal of Peasant Studies* 36, No. 1, 227–243. DOI: 10.1080/03066150902820487.

Lefebvre, H. 2002. Comments on a new state form. *Antipode* 33, No. 5, 769–782. DOI: 10.1111/1467-8330.00216.

Mamdani, M. 1996. *Citizen and subject. Contemporary Africa and the legacy of late colonialism.* Princeton, N.J., Chichester: Princeton University Press (Princeton Studies in Culture/Power/History).

Marcuse, P. 2009. From critical urban theory to the right to the city. *City: Analysis of Urban Trends, Culture, Theory, Policy, Action* 13, No. 2-3, 185–196. DOI: 10.1080/13604810902982177.

McClintock, N. 2014. Radical, reformist, and garden-variety neoliberal: coming to terms with urban agriculture's contradictions. *Local Environment* 19, No. 2, 147–171. DOI: 10.1080/13549839.2012.752797.

McMichael, P. 2016. Commentary. Food regime for thought. *The Journal of Peasant Studies* 43, No. 3, 648–670. DOI: 10.1080/03066150.2016.1143816.

Montenegro de Wit, M., Shattuck, A., Iles, A., Graddy-Lovelace, G., Roman-Alcalá, A., & Chappell, M. J. 2021. Operating principles for collective scholar-activism: Early insights from the agroecology research-action collective. *Journal of Agriculture, Food Systems, and Community Development*. Advance online publication. DOI: 10.5304/jafscd.2021.102.022.

O'Brien, K.J. 1996. Rightful resistance. *World Pol.* 49, No. 1, 31–55.

O'Brien, K.J. 2013. Rightful resistance revisited. *Journal of Peasant Studies* 40, No. 6, 1051–1062. DOI: 10.1080/03066150.2013.821466.

Purcell, M. 2002. Excavating Lefebvre: The right to the city and its urban politics of the inhabitant. *GeoJournal* 58, No. 2/3, 99–108. DOI: 10.1023/B:GEJO.0000010829.62237.8f.

Purcell, M. 2014. Possible worlds. Henri Lefebvre and the right to the city. *Journal of Urban Affairs* 36, No. 1, 141–154.

Purushothaman, S. and Patil, S. 2019. Agrarian change and urbanization in Southern India. *City and the Peasant*. Springer: Singapore. DOI: 10.1007/978-981-10-8336-5.

Pye, O. 2010. The biofuel connection – transnational activism and the palm oil boom. *The Journal of Peasant Studies* 37, No. 4, 851–874. DOI: 10.1080/03066150.2010.512461.

Sage, C., Kropp, C. and Antoni-Komar, I. 2020. Grassroots initiatives in food system transformation: the role of food movements in the second "Great Transformation". In Kropp, C., Antoni-Komar, I., Sage, C. (Eds): *Food systems transformations. Social movements, local economies, collaborative networks*. New York: Routledge, 1–19.

Sbicca, J. 2014. The need to feed. Urban metabolic struggles of actually existing radical projects. *Critical Sociology* 40, No. 6, 817–834. DOI: 10.1177/0896920513497375.

Scott, J. C. 1985. *Weapons of the weak. Everyday forms of peasant resistance*. Princeton, N.J.: Yale University Press.

Shivji, I.G. 2017. The Concept of "Working People". *Agrarian South: Journal of Political Economy* 6, No. 1, 1–13. DOI: 10.1177/2277976017721318.

Sibhatu, K.T. and Qaim, M. 2017. Rural food security, subsistence agriculture, and seasonality. *PLoS One* 12. No. 10, DOI: 0186406.

Siebert, A.; May, J. 2016. Urbane Landwirtschaft und das Recht auf Stadt. Theoretische Reflektion und ein Praxisbeispiel aus George, Südafrika. In Engler, S., Stengel, O., Bommert, W. (Eds.): *Regional, innovativ und gesund. Nachhaltige Ernährung als Teil der Großen Transformation*. Göttingen: Vandenhoeck & Ruprecht, 153–168.

Skinner, C. 2018. Contributing yet excluded?: informal food retail in African cities. In Battersby, J., Watson, V. (Eds.): *Urban food systems governance and poverty in African Cities*. New York: Routledge, 2018. | Series: Routledge studies in food, society and the environment: Routledge, 104–115.

Skinner, C.; Haysom, G. 2017. The informal sector's role in food security. A missing link in policy debate. *Hungry Cities Partnership - Discussion Papers*, No. 6. http://hungrycities.net/wp-content/uploads/2017/03/HCP6.pdf, Accessed on 09/04/2021.

South African Food Sovereignty Campaign 2015. Report. Johannesburg. https://safsc.org.za/wp-content/uploads/2014/03/FS-Assembly-Report.pdf, Accessed on 10/12/2020.

Tapscott, C. 2010. Social mobilization in Cape Town: a tale of two communities. In Thompson, L., Tapscott, C. (Eds.): *Citizenship and Social Movements perspectives from the global South*. London: Zed Books, 260–278.

Tornaghi, C. 2017. Urban agriculture in the food-disabling city: (Re)defining urban food justice, reimagining a politics of empowerment. *Antipode* 49, No. 3, 781–801. DOI: 10.1111/anti.12291.

United Nations Population Division 2019. World Urbanization Prospects: 2019. https:// data.worldbank.org/indicator/SP.URB.TOTL.IN.ZS?locations=ZA, Accessed on 10/04/21.

Visser, O.; Mamonova, N.; Spoor, M.; Nikulin, A. 2015. "Quiet Food Sovereignty" as food sovereignty without a movement? Insights from Postsocialist Russia. *Globalizations* 12, No. 4, 513–528. DOI: 10.1080/14747731.2015.1005968.

Walker, C. 2007. Redistributive land reform: for what and for whom? In Ntsebeza, L., Hall, R. (Eds.): *The Land Question in South Africa. The challenge of transformation and redistribution.* Cape Town: HSRC Press, 132–151.

Wittman, H. 2009a. Reframing agrarian citizenship. Land, life and power in Brazil. *Journal of Rural Studies* 25, 120–130. DOI: 10.1016/j.jrurstud.2008.07.002.

Wittman, H. 2009b. Reworking the metabolic rift. La Vía Campesina, agrarian citizenship, and food sovereignty. *The Journal of Peasant Studies* 36, No. 4, 805–826. DOI: 10.1080/03066150903353991.

Wright, E.O. 2017. How to be an anti-capitalist for the 21st century. *THEOMAI Journal Critical Studies about Society and Development* 35, No. 1, 8–21.

Wright, E.O. 2019. *How to be an anticapitalist in the 21st century.* London: Verso.

Zhan, S.; Scully, B. 2018. From South Africa to China. Land, migrant labor and the semi-proletarian thesis revisited. *The Journal of Peasant Studies* 45, No. 5-6, 1018–1038. DOI: 10.1080/03066150.2018.1474458.

6 Conclusions
Urban South Africa and beyond

This book illuminates food sovereignty in practice through a comprehensive analysis of the urban agriculture initiative KEF in the medium-sized city of George, specifically food producers' lived realities, struggles, and resistance in an exclusionary food landscape. In an attempt to make nutritious food for all an everyday reality, members engage in a range of food cultivation practices and sharing networks. These experiences advance the food sovereignty discourse to comprise urban and peri-urban settings. So far, closer investigations of the realities of urban food producers and their political engagement remain rare in the Global South. Mounting pressures of the commercialised agri-food system – ranging from rising food prices to malnutrition – served as the point of the departure and were illuminated at the beginning of the book. In urbanised South Africa, many people are subject to these dire conditions which are further complicated by insufficient integration in the labour market and historical inequalities. Particularly at the time of the COVID-19 pandemic, the role of urban agriculture in advancing food and nutrition security has been considered critically important (cf. Lal 2020). In addition, alternative food provision like community soup kitchens and food parcels regained relevance in many places around the world. Clearly, the disparities of the prevailing system become ever more visible under the burning lens of the pandemic.

Considering such developments, the conclusion of this book refers back to its theoretical contributions in particular the critical urban food perspective and the related key findings. Using dynamics of the globalised agri-food system as an overall framing while situating this research in a specific urban context in the Global South, the book's analytical perspective is both globally oriented and contextually grounded. Certainly, the innovative analytical perspective and its possible applicability elsewhere is one of the key contributions and is briefly outlined in the remaining part. Direct findings based on this perspective are considered as three further contributions. First, the book provides novel insights into food producers' lived realities and struggles at the urban outskirts. Second, the political dimension of urban agriculture initiatives and food producers' agency are illuminated. Finally, and closely linked to the preceding aspects, this work elucidated potential alliances theoretically to overcome urban agriculture's niche existence and to propel related food sovereignty efforts. So overall, this chapter brings together the various stands of this research in a final discussion and points out more over-arching reflections

DOI: 10.4324/9781003182634-6

that draw together empirical findings with theoretical insights. Ultimately, it shows what others can take and advance from the specific analytical perspective and case as well as its wider appeal for different locations.

Critical urban food perspective

The book introduces an innovative analytical lens, the critical urban food perspective – to explore exclusionary dynamics in everyday life from the viewpoint of the deprived in the dominant agri-food system. Directly related to that, this work has a strong focus on people's responses and proposed alternatives. In general, this approach applied a constructive triad consisting of analysis, critique, and search for alternatives chiming with Peter Marcuse's notion of exposing inequalities, proposing alternatives, and politicising this work as part of the right to the city (2009). It is this basic interplay of critical observation and appreciation of emerging opportunities which makes this perspective relevant to those engaging with social movement responses to restricted food and land access.

The intention of this analytical perspective is to shed light on the lives of marginalised urban dwellers, a group of social actors which has been barely considered in the food sovereignty discourse thus far. This book focuses specifically on backyard food producers and squatter farmers at the outskirts of cities as well as their struggles over food and land. To learn from their experience, the developed perspective navigates through vast fields of research which engage with related urban and rural particularities. Hence, this lens builds upon rich knowledge and concepts provided by critical urban theory and critical agrarian studies. In this context, the book provides a fresh perspective on several divides and labels evident in both schools of thought, for instance traditional/modern city, rural agrarian/urban modern society, labourer/smallholder, and de-peasantisation/re-peasantisation (e.g. Bernstein 2011; Holt-Giménez and Shattuck 2011; Lefebvre 2003 [1970]). For instance, several food producers are not fully integrated in the urban labour market and partly rely on agrarian livelihoods. Here, questions of land and livelihood go beyond an artificial rural–urban divide. Consequently, strict divides are not always possible, nor are they necessarily helpful. Thus, existing categories need to be adapted to lived experiences, and consideration needs to be given to the spaces between, in order to understand today's urban realities which can barely be separated from those of the countryside. A key message is that reshaping and broadening theoretical boundaries are essential in exploring new actor constellations, mobilisations, and ways of representation.

Critical agrarian studies help in exploring the prevailing agri-food system. In South Africa, scholars observed a "stalled agrarian transition" intertwined with dynamics of "semi-proletarianisation" (Du Toit 2018; Cousins et al. 2018). These considerations are essential if we are to understand the lived realities and means of survival of urban food producers in George. The combination of different theoretical perspectives is useful beyond South Africa, particularly in ever-growing (peri-) urban settings in which deprived dwellers (partly) rely on agrarian livelihoods, critically engage with and respond to the restrictive agri-food system in place. In such contexts, critical urban theory particularly Marcuse's (2009) analytical steps of the right to the city are considered ideal in structuring and understanding transformative efforts of marginalised food producers. The notion of everyday life serves as the point of departure and is to be complemented by diverse notions of resistance and space. The concepts of everyday forms of peasant resistance and everyday politics connect with critical urban theory and help to pinpoint out urban food producers' daily practices and related claims (cf. James Scott, Benedict Kerkvliet, Kevin O'Brien). Diverse forms of resistance and compliance are considered as integral elements to challenge a system characterised by domination and poverty (Scott 1985, 289). Related theoretical and analytical considerations offer a frame which is sensitive to lived experiences on the ground. This is a decisive pre-condition for research on food sovereignty. It must be highlighted that the application of this analytical framework benefitted greatly from rich qualitative research methods mainly participatory in nature, which allowed to explore diverse facets of everyday politics and people's perceptions.

In sum, the introduced critical urban food perspective and related guidance can be applied beyond the boundaries of South Africa. For instance, migrants' food cultivation in so-called international gardens or allotments in different towns of Germany seems to offer an unexplored and incipient breeding ground for food sovereignty. Many of them grow crops which are otherwise not readily available and sell them for an additional income. The combined theoretical concepts could also be helpful in analysing resistance and cooperation of marginalised food growers in urbanised Brazil, in which alliances between different movements are more established and often connected to bigger food sovereignty movements like the Brazilian Landless People's Movement (Movimento dos Trabalhadores Rurais Sem Terra).

Moreover, it seems essential to develop the critical urban food perspective further to engage explicitly with further actors and their interactions with civil society. For instance, social movements have been working in and outside, through and between structures of the state. A critical urban food perspective may add new insights on different ways of using state power to advance efforts towards food sovereignty in urban settings. This notion could benefit from literatures on legal rights and social justice. In this context, further methodological and empirical work would greatly add to the existing perspective.

Lived realities of food producers at the urban margins

The critical urban food perspective navigates through the deprived realities of urban and peri-urban food producers in terms of food sovereignty including issues of resource access and control. So far, related experience from the fringes of cities, particularly secondary cities in the Global South, are barely available. In the intermediate city of George, the researched neighbourhoods showed a prevailing rural character in structure and social interactions. At the same time, close ties to the city centre shape the food retail system and tend to render local, informal food purchase structures redundant. The findings revealed that the KEF initiative benefits from several rural characteristics of this smaller city. For instance, different neighbourhoods are within reach, which enables exchange. Residents know the government officials and thus are in the position to make contact and target their interventions directly. Some of the communities were separate villages in the past and thus have a long history and value traditions as well as local knowledge. Moreover, many inhabitants possess ruptured ties to the countryside, and their families have been growing food for generations. However, other characteristics show similarities with metropolitan areas, for instance, the overcrowded and densely built township area on the outskirts of the city.

The book shows that many of the food producers are not fully integrated in the labour market and struggle to meet their dietary needs. For instance, precarious employment and low wages require people to engage in multiple income strategies and to keep costs for social production low which mirrors Shivji's notion of the working people (2017). In this context, the critical urban food perspective illuminated that the role of food access and choices as well as the land for food production become more important which reveals parallels to and the importance of (urban) agrarian livelihoods. These two outstanding dimensions of food and the land and related findings are outlined briefly in the following paragraphs.

In the context of food, the initiative's lived experiences show that the dominant food system restricts people's food choices and challenges them to create alternatives. People on low wages are forced to purchase cheap food. In addition, supermarkets and the groceries on offer do not necessarily meet the needs of the poor for instance in terms of package size and payment method. With a strong corporate turn, the state seems unable to protect them from these inequalities. In addition, the commercialised retail landscape and stressful workdays have changed dietary patterns with an increase in cheap processed and convenience food. Fresh meals, fruits, and vegetables seem less affordable. Many KEF members grapple with malnutrition and are in need of healthier diets. Some members have been concerned about these developments and moreover are critical of the use of antibiotics in meat production, and genetically modified seeds. These difficulties are the driving forces of the initiative's interventions in urban agriculture, which comprise the promotion of several kinds of food production, ranging from backyard gardens to squatter farming. Members have established a platform to share knowledge, food, seeds, and tools. Some even sell their produce informally. Several supporters and donors – of material donations in the form of seeds, seedlings, or trees – encourage KEF's work. Taken together, this can be considered an attempt to create a food system that benefits the many not just the few. Some members even dream of independence from the industrial agri-food system, with self-sufficiency the ultimate goal. Although these attempts sound promising, reliance on additional staples such as cooking oil from grocery stores seems difficult to avoid. It is striking that KEF rejected the development interventions of the municipality, with members criticising the municipality for a lack of sensitivity to the issues on the ground. The inadequate food security interventions of the George Municipality and the Department of Agriculture, which are mainly framed within urban agriculture projects, mirror the government's failure to engage in broader and more systemic interventions against malnutrition.

Land is another dominant feature in the initiative's work. Despite the presence of fallow land between different neighbourhoods and at the fringes of the city, KEF has been struggling to gain official access in different sites. Farmers in Thembalethu have occupied land and have requested to be integrated in governmental development programmes. Despite the government's Proactive Land Acquisition Strategy, the promotion of urban commonage country-wide, and the proposals of integrated development plans, these squatters have not been able to benefit and have been waiting for support from the government for a long time. The case of the Thembalethu farmers exemplifies how the government maintains the precarious status quo and even adds further to marginalisation. Similar cases, including land occupations at the fringes of cities and shifting governmental promises, have been observed across the country and some of them are well documented, e.g. at the fringes of Khayelitsha, Cape Town (Jacobs 2018) and in Kennedy Road, Durban (Patel 2011). In other communities, in Blanco and Pacaltsdorp, the initiative requested official land access for community gardens in the former buffer zone and fallow municipal farmland. The location of these pieces of land presented an opportunity to connect neighbourhoods which were separated under the apartheid regime. However, access was denied. Although the government's role is to balance diverse urban land use purposes, housing remains in the spotlight of urban land use planning. Local authorities seem to be tempted to shift their focus to profitable real-estate investments. This distortion exemplifies hierarchical planning mechanisms, which do not provide entry points for the people, but deepen historical inequalities and fuel public discontent. Against this background, the lens of the critical urban food perspective sheds light on the rather neglected urban dimension of the resurgent land debate in South Africa.

Overall, this reliance on the land and difficulties in self-provision vividly expose the cracks in the system and various challenges in making land-based livelihoods secure amidst intertwined dynamics of exclusion, precariousness, informality, and un(der)employment. Based on these insights and in terms of the bigger picture, the book does not only demonstrate that urban agriculture proliferates in marginalised urban settings, but it also clearly shows that people on low wages and in precarious employment are particularly affected by an increasing crisis of social reproduction. In fact, in such contexts urban agriculture can be considered an expression of the roll-back of the state and implies a "neoliberal ethos of self-responsibilization" (Rosol 2012, 251). These outlined dire conditions become ever more visible globally particularly over the course of multiple unfolding crises referring, for instance, to recurring and intensifying climate crisis-related volatile food prices and water shortages as well as the COVID-19 pandemic. Hence, a key take-away message for those promoting and celebrating urban agriculture as self-help strategy for the poor is to consider the harsh realities in which food cultivation occurs and which often restrict farming in the city. Ideally, the provided insights encourage other researchers and social movements to consider, illuminate, and support these manifold struggles of those growing food under challenging conditions. Part of it is the larger socio-economic structure that deprives people, not only of a voice but also of stable labour market integration, fair wages, proper food, education, and healthcare. This has underscored the urgency of the overarching task of food sovereignty construction. Drawing insights from a powerful urban movement shows that addressing food and land inequality on a local scale is essential and requires work on multiple fronts and engagement with different actors for instance the state, as highlighted previously.

Political dimensions and food producers' agency

Guided by the critical urban food perspective, this book provides important insights on urban agriculture's political dimension and food producers' agency. Hence, it encourages others to consider and advance these often underestimated resistances in South Africa and elsewhere. The findings of the case study show that so-called external development interventions, for instance by the municipality in promoting urban agriculture to improve food security or income creation, have a tendency to be rather short term and to ignore wider systemic inequalities, for instance precarious land or market access. Thus, the book makes an essential contribution in zooming in on food producers' demands and alternatives they already created. According to the right to the city, urban food producers represent their needs through different ways of reconfiguring urban space. Growing food can therefore challenge taken-for-granted functions and the use of space can be political (Wekerle and Classens 2015, 1175). Chapter 5 highlights how KEF signals the need for transformative food and land politics as well as citizen-based urban renewal. Members' garden and farm plots are perceived as spaces of contestation, land becomes a means of negotiation, and food production exposes and expresses critique of unjust and uneven development (cf. Eizenberg 2012, 767 and 775).

The investigated experiences and claims were framed as eight interrelated rights to the city, which comprise the political implications of the movement (see Chapter 5): (1) right to access quality and locally grown food, (2) right to control own diets, (3) right to have a determining influence on local retail environment, (4) right to have a determining influence on governmental community interventions, (5) right to access cultivable land, (6) right to an enabling knowledge environment, (7) right to environmental conservation, and (8) right to solidarity. Following the lead of the food sovereignty concept, these rights can be seen as a proposal and socio-political goals for active participation in decision-making beyond a passive role in government programmes and for further politicisation. Still, these proposed rights are not legally enforceable and are rather means for further strategic and political work. Hence, the vivid efforts of initiatives like KEF provide starting points for local governments' interventions in resolving issues of hunger, malnutrition, market access, and mitigating the impacts of climate change to name just a few. According to O'Brien's (2013) notion of rightful resistance, vulnerable people's proposals can spark new dialogues, strengthen social identities, pose troubling questions, and spur desire for political change. Acknowledging these diverse claims and interventions through the critical urban food perspective furthers the reach of the academic food sovereignty discourse and movement. It is thus essential that food production in cities overcomes its status of a marginalised activity.

Beyond transformative attempts of such an initiative, urban agriculture can always be at risk of policies of containment and co-option. Many scholars have done important work in revealing that urban agriculture in the Global North, including its spaces of self-organisation and community volunteering, is embedded in and constrained by the exclusionary dynamics of neoliberal cities and the capitalist logics of exploitation (cf. Certomà and Tornaghi 2015; Roman-Alcalá 2018). Some of these dynamics are evident in the KEF initiative. For instance, the municipality wanted to use the initiative's efforts in urban agriculture to create measurable results for its own community development plans. However, the needs and wishes of the participants were not explicitly considered, which caused the initiative to reject governmental interventions. In such contexts, Gliessman, Friedman, and Howard observe that "an ongoing dance of creativity and appropriation exists between grass-roots inventiveness and corporate and government co-optation" (2019, 104). While they refer to the transformative potential of this dance, they highlight the role of food sovereignty in steering this process by focusing on related explicit aims like appropriate land access. Consequently, research on and practices of these localised alternatives must constantly ask if these efforts follow the intended direction.

Building from the empirical findings provided in Chapters 3–5 illuminates important messages for further politicisation of local food initiatives and urban food governance. While research on urban food systems in Southern Africa has pointed in different ways to the importance of (participatory) urban food policies, governance, and planning, there has been a shortage of practical implementation (Battersby and Haysom 2018). It is important to step beyond hierarchical power structures and the sole focus on economic development. Based on the experience of the KEF movement, top-down development interventions seem inadequate to address problems of hunger, malnutrition, and land access. At the same time, the corporate sector's role in shaping urban retail planning is considered too powerful. In this regard, urban policies neglect the important role of the informal sector in food provision and (often the only) point of sale for produce grown by small local producers. It is hence essential to provide democratic structures in food system planning and governance which favour those who inhabit the city. In this context, democratic multi-stakeholder platforms, so-called food policy councils strongly driven by civil society, are becoming more popular globally. Clearly, their potential, work, and impacts require further research particularly in the Global South. The final section of this chapter briefly illuminates this point.

Trajectories of food sovereignty and alliances

A fourth contribution coming out of this book refers to importance of creating wider visibility of and meaningful alliances for food sovereignty. In fact, there is no singular conception of food sovereignty but rather multiple, overlapping efforts. It is one contribution of the critical urban food perspective to bring together and appreciate these efforts, diverse forms of resistance, and alternatives within the prevailing agri-food system. Hence, further work on these emergent pathways and related knowledge creation is urgently required to advance food sovereignty in discourse and practice, in and outside of cities.

Although the food sovereignty discourse is rather nascent on national level in South Africa, it is important to keep in mind that food sovereignty trajectories look different at various scales and in different settings (Schiavoni 2017). Despite the decline of small-scale farming, particularly in the apartheid era, food sovereignty efforts are not entirely absent as outlined in Chapter 2. Adjusting the lens to local case-specific particularities reveals different forms of resistance against the prevailing commercialised agri-food system. In this context, it is important to stress out that food sovereignty's proliferation is not limited to a linear trajectory: efforts on the global or national level do influence grassroots struggles. At the same time, as emphasised with the critical urban food perspective, initiatives on the ground could present a breeding ground for food sovereignty and initiate wider struggles.

Food sovereignty experiences in different places show that it is essential to build and maintain relationships with the broader civil society, institutions, landscapes, and ecosystems. Although a food sovereignty campaign and broader discourse is taking shape at the national level in South Africa, it lacks stronger consideration and articulation of the voices on the ground. The book reveals shortcomings in this regard, for instance smallholders' agency and the broader critical urban food community seem invisible in well-articulated food sovereignty demands and movements. For instance, urban agriculture groups, community gardeners, or organised livestock farmers in townships which seek to improve the situation of their deprived communities might even be interested in revealing larger inequalities. At the same time, the eclectic landscape of alternative urban- and peri-urban farming, basically niches within the exclusionary food system, is highly fragmented and difficult to grasp from outside. Certainly, localised groups and their members do not necessarily have the resources to engage in interventions beyond their daily practices of food production, such as taking part in public protest, engaging regularly in municipal meetings, or networking activities beyond their community. The embeddedness of these localised efforts in the prevailing capitalist social system and the dominant agri-food system underlines the importance of strong alliances within civil society including support of groundswell initiatives. Simultaneously, meaningful interaction with the state and market is necessary. This is in line with Wright's thoughts on eroding capitalism on multiple fronts (2019).

This experience is not unique to South Africa. In fact, voices from below in local settings are often not directly on the radar of established civil society organisations. When identifying social movements and organisations that struggle against inequality, practitioners and scholars alike tend to focus on high-profile adversarial social movements and campaigns. Unique histories, resistances, and hidden situations on the ground in everyday life tend to be neglected, thereby missing an important aspect of potential food sovereignty politics. Particularly transnational food sovereignty movements and discourses are likely to be at risk of losing political momentum when it comes to local issues (Edelman and Borras 2016). While food sovereignty calls for a stronger consideration and articulation of the voices on the ground, it also calls for alliances. Exchange and maybe even joint strategies are essential. Although KEF has already advanced its local network, connections and exchange with other initiatives might further politicise their work, contribute to explicit strategies in their daily struggles, or expand their work to rural areas. Following the intention of the critical urban food perspective, alliances and common interventions for food sovereignty should unite people and issues in urban and rural agrarian settings. Beyond food sovereignty, it is crucially important to explore further strategic coalitions with social movements or advocacy networks that for instance fight labour precarity or environmental degradation (Brenner 2019, 392).

Chapter 5 particularly highlights the role of possible alliances, the combination of radical and progressive transformative interventions, and the joint tasks in Wright's anti-capitalist strategies (2019). In many different places around the world, attempts to strengthen connections between different actors and places, both urban and rural, are already underway and should be understood as inspiration for practical and political work elsewhere. International experiences show that food policy councils have been able to unite different actors to improve and fight for alternative, social, and ecologically just local food systems. Localised successes have been celebrated for instance in cities in Brazil, Canada, Germany, Madagascar, and the USA. These interventions have parallels with food sovereignty's intention to restore a discourse of public control of food systems and push for new political spaces and spaces for engagement. Direct producer–consumer, rural–urban linkages, collective purchasing groups (food cooperatives), and direct purchases by government institutions have recently (re)gained attention and helped to avoid some of the challenges related to a global and commercialised agri-food system. For instance, the Hansalim Cooperative in Seoul, South Korea, connects about 2,000 rural small-holders with about 1.5 million urban consumers (Thurn, Oertel, and Pohl 2018). The initiative is motivated by attempts at self-sufficiency, import independence, and stable prices for producers and consumers. The critical urban food perspective is considered helpful to explore such innovative coalitions and types of organisations as well as their concrete interventions under the banner of food sovereignty in different places around the world. Overall, these promising networks and alliances as well as their progressive efforts require specific attention in future work of grassroots organisations, academics, and all those working towards more just agrifood systems.

References

Battersby, J.; Haysom, G. 2018. Linking urban food security, urban food systems, poverty, and urbanisation. In Battersby, J., Watson, V. (Eds.): *Urban Food Systems Governance and Poverty in African Cities*. New York: Routledge, 2018. | Series: Routledge studies in food, society and the environment: Routledge, 56–67.

Bernstein, H. 2011. 'Farewells to the peasantry?' and its relevance to recent South African debates. *Transformation* 75, No. 1, 44–52.

Brenner, N. 2019. *New urban spaces: Urban theory and the scale question*. New York: Oxford University Press.

Certomà, C.; Tornaghi, C. 2015. Political gardening. Transforming cities and political agency. *Local Environment* 20, No. 10, 1123–1131. DOI: 10.1080/13549839.2015.1053724.

Cousins, B.; Dubb, A.; Hornby, D.; Mtero, F. 2018. Social reproduction of "classes of labour" in the rural areas of South Africa. Contradictions and contestations. *The Journal of Peasant Studies* 45, No. 5-6, 1060–1085. DOI: 10.1080/03066150.2018.1482876.

Du Toit, A. 2018. Without the blanket of the land. Agrarian change and biopolitics in post–Apartheid South Africa. *The Journal of Peasant Studies* 45, No. 5-6, 1086–1107. DOI: 10.1080/03066150.2018.1518320.

Edelman, M.; Borras, S.M. 2016. *Political dynamics of transnational agrarian movements*. Rugby: Practical Action Publishing.

Eizenberg, E. 2012. Actually existing commons. Three moments of space of community gardens in New York City. *Antipode* 44, No. 3, 764–782. DOI: 10.1111/j.1467-8330.2011. 00892.x.

Gliessman, S.; Friedmann, H.; Howard, H. 2019. Agroecology and food sovereignty. *IDS Bulletin* 50, No. 2.

Holt-Giménez, E.; Shattuck, A. 2011. Food crises, food regimes and food movements: Rumblings of reform or tides of transformation? *The Journal of Peasant Studies* 38, No. 1, 109–144. DOI: 10.1080/03066150.2010.538578.

Jacobs, R. 2018. An urban proletariat with peasant characteristics: Land occupations and livestock raising in South Africa, The Journal of Peasant Studies 45, No. 5-6, 884–903. DOI: 10.1080/03066150.2017.1312354.

Lal, R. 2020. Home gardening and urban agriculture for advancing food and nutritional security in response to the COVID-19 pandemic. *Food Sec.* 12, 871–876. DOI: 10.1007/ s12571-020-01058-3.

Lefebvre, H. 2003 [1970]. The Urban Revolution. Minneapolis: University of Minnesota Press.

Marcuse, P. 2009. From critical urban theory to the right to the city. *City: Analysis of Urban Trends, Culture, Theory, Policy, Action* 13, No. 2-3, 185–196. DOI: 10.1080/13604810902982177.

O'Brien, K.J. 2013. Rightful resistance revisited. *Journal of Peasant Studies* 40, No. 6, 1051–1062. DOI: 10.1080/03066150.2013.821466.

Patel, R. 2011. Fairytale violence or Sondheim on solidarity, from Karnataka to Kennedy Road. In Essof, S., Moshenberg, D. (Eds.): *Searching for South Africa. The new calculus of dignity*. Pretoria: Unisa Press, 190–220.

Roman-Alcalá, A. 2018. (Relative) autonomism, policy currents and the politics of mobilisation for food sovereignty in the United States. The case of Occupy the Farm. *Local Environment* 23, No. 6, 619–634. DOI: 10.1080/13549839.2018.1456516.

Rosol, M. 2012. Community volunteering as neoliberal strategy? Green space production in Berlin. *Antipode* 44, No. 1, 239–257. DOI: 10.1111/j.1467-8330.2011.00861.x.

Schiavoni, C.M. 2017. The contested terrain of food sovereignty construction. Toward a historical, relational and interactive approach. *The Journal of Peasant Studies* 44, No. 1, 1–32. 10.1080/03066150.2016.1234455.

Scott, J. C. 1985. *Weapons of the weak. Everyday forms of peasant resistance*. Princeton, N.J.: Yale University Press.

Shivji, I.G. 2017. The Concept of "Working People". *Agrarian South: Journal of Political Economy* 6, No. 1, 1–13. DOI: 10.1177/2277976017721318.

Thurn, V.; Oertel, G.; Pohl, C. 2018. *Genial lokal. So kommt die Ernährungswende in Bewegung*. München: oekom.

Wekerle, G.R.; Classens, M. 2015. Food production in the city. (Re)negotiating land, food and property. *Local Environment* 20, No. 10, 1175–1193. DOI: 10.1080/13549839.2015.1007121.

Wright, E.O. 2019. *How to be an anticapitalist in the 21st century*. London: Verso.

Index

Page numbers followed by 'n' refer to notes numbers.

Printed in the United States
by Baker & Taylor Publisher Services